15 March 2010 - aetatis meae xxxi - Countryside - $2 -

Storms in Space

Storms in Space is the story of the exciting and mysterious region between the Earth and the Sun, where violent storms rage unseen by human eyes. Disruption of spacecraft and satellites, television transmission failures, and power blackouts are just a few of the effects of this powerful force of nature, caused by the charged particles and electromagnetic force fields that dominate space.

This is a highly readable synopsis of man's current understanding of the space environment. The book discusses the strong similarities between terrestrial weather and space weather, and goes on to describe the causes and effects of space storms, and how they can be monitored by satellites and from observatories on Earth. The forecasting of storms in space is presented, along with prospects for improved models in the future.

Freeman's compelling story is written in a clear and engaging fashion, accessible to all levels of readership. The book will be valuable to space scientists, physicists, astronomers and anyone with an interest in understanding the phenomenon of space weather.

JOHN FREEMAN is Professor Emeritus and Research Professor of Physics and Astronomy at Rice University, Houston, Texas. His primary research interests include computer modeling of the Earth's magnetosphere for space weather specification and prediction, neural networks and other artificial intelligence applications to forecasting geophysical parameters.

Storms in Space

John W. Freeman
Foreword by George Siscoe

CAMBRIDGE
UNIVERSITY PRESS

PUBLISHED BY THE PRESS SYNDICATE OF THE UNIVERSITY OF CAMBRIDGE
The Pitt Building, Trumpington Street, Cambridge, United Kingdom

CAMBRIDGE UNIVERSITY PRESS
The Edinburgh Building, Cambridge CB2 2RU, UK
40 West 20th Street, New York, NY 10011–4211, USA
10 Stamford Road, Oakleigh, VIC 3166, Australia
Ruiz de Alarcón 13, 28014 Madrid, Spain
Dock House, The Waterfront, Cape Town 8001, South Africa

http://www.cambridge.org

First published 2001

Printed in the United Kingdom at the University Press, Cambridge

Typeface Trump Mediaeval 9.5/15 pt.   *System* QuarkXpress™ [HMCL]

A catalogue record for this book is available from the British Library

Library of Congress Cataloguing in Publication data

Freeman, John W.
Storms in space/John W. Freeman; foreword by George Siscoe.
    p.   cm
Includes bibliographical references and index.
ISBN 0 521 66038 6
1. Space environment. I. Title.
QB505 .E74 2001
629.4′16–dc21   2001025569

ISBN 0 521 66038 6 hardback

Contents

Foreword

John Freeman has done the impossible, written a book on space weather that's a fun read. It's hard enough to create a non-fiction book on regular weather that's a fun read, unless like Sebastian Junger's *The Perfect Storm* it's about people and violence. The genius of John's book is its inventiveness. In it we read a short story that *is* about people and violence and that tells us what space storms are and what they do. We take a whirlwind tour through Earth's space weather machine, called the magnetosphere, riding a proton then an electron. We listen to an interview with Joe Allen, the world's premier archivist of space-weather damage who knows where the bodies are buried. The important facts are here, told by an insider with a talent for telling the significant and discarding the rest.

John's own contribution to space weather will go in history books when the subject grows old enough to have a history. Working with colleagues at Rice University, he developed the first numerical space-weather model to go into operation at a national space weather center. A corresponding event happened in the history of meteorology in 1955, when numerical weather forecasting was inaugurated at the US Weather Bureau (as it was called then). That event led over time to the present highly successful forecasting capability of the National Weather Service (jokes to the contrary notwithstanding). John and his co-workers have put space weather on the same path that, because numerical codes constantly improve through incorporating better physics and better data, leads inevitably to steady improvement.

This book is a masterly blend of the people side and the technical side of space weather. I wish I had written it.

George Siscoe

Dedication

This book is dedicated to Alexander J. Dessler whose boundless enthusiasm, friendship and insight have inspired a generation of scientists, and also to my loving wife who has been my inspiration for 45 years.

Preface

The purpose of this book is to provide a brief glimpse of an astounding and beautiful aspect of nature known only to a relative handful of space scientists and yet which is capable of disrupting human technical systems ranging from communication satellites to electric power grids—storms in space.

The problem faced by a specialist writing for a general audience encompasses several challenges: first, he or she must draw connections to everyday experience that lead the reader into the realm of the new phenomenon along familiar and enticing paths; second, the writer must avoid technical jargon and translate the descriptions into understandable word images; and last, the unveiling of new concepts must be layered and progressive as the reader's comprehension matures.

The problem of describing storms in space is doubly challenging because the majority of the phenomenon is invisible to the human eye. This is why little progress could be made studying the phenomenon until the space age when in-situ satellite investigations became possible. It is also why we space physicists are continually jealous of astronomers who can show marvelous images of the objects of their studies.

I have undertaken these challenges, partly because of a desire to explain to my friends what my work is all about, but in a larger sense because I want to share with students and the public some of the awesome beauty of the unseen world of space. A further objective is to foster a general appreciation of the hazards that storms in space represent in order that they may gain some respect as a significant force in nature.

One approach to the first challenge is to draw parallels between storms in space and storms on the Earth. As we shall see, these parallels are numerous and instructive. At the same time, the differences lead us to a description of elements of nature that may be new to us. The physical elements that make great storms on the Earth are well known to us. The wind and rain, clouds, thunder and lightning, snow, ice and hail all leave no doubt about the presence or imminence of a storm. Even though we can't see the wind we feel its effect on our skin and our clothing, and we see trees sway.

In stark contrast, the elements of a storm in space are unfelt by the human senses. These elements are subatomic charged particles and force fields detectable only by charged particles as they move freely in the vacuum of space. They form a very low-density, almost ethereal, and essentially invisible medium that pervades the vast space near the Earth and beyond. It is somewhat of an enigma that out of this seemingly insignificant medium can grow great storms in space that are capable of wreaking havoc on human systems.

This book is dedicated to revealing the nature of these storms and some of their awesome and dangerous effects.

John W. Freeman

The Cast of Characters

The elements of storms in space are fundamental charged particles and electromagnetic (force) fields. Together these two physical quantities make up the cast of characters that are purveyors of the tempests in space. It will help our appreciation of the storms if we take a few minutes to meet these characters.

Charged particles come in two varieties, negatively charged and positively charged particles. The most stable of the negative particles is the famous electron. As the negative, orbital component of all atoms, electrons are everywhere. In another guise, electrons have become the workhorses of the human race. Coursing through electrical wires, they are the essential ingredients in practically all of our modern technology. In space, where electrons can be free from their atomic bondage and there are no wires to guide them, the motion of electrons is more complex and interesting. Uninhibited by frequent collisions with atoms, they become capable of achieving high speeds, close to the speed of light. Speed means kinetic energy. Kinetic energy is the status symbol for charged particles in space and is of great importance to storms in space. Electrons are extremely small and have very little mass. To achieve a certain kinetic energy they must make up for their low mass by high velocities.

The other main variety of charged elementary particle, of course, is the positive particles. The most famous of the positive particles is the proton. Protons make up the charged component of the nucleus of atoms and are about 2,000 times more massive than the orbiting electrons. But, like electrons, protons in space can exist as isolated, independent particles where they also are subject to complex motion produced by force fields. In addition to protons, we also have ions as positively charged particles. Ion is the generic name for an atom that has lost one or more of its orbital electrons to obtain a net positive charge. Ions may move and exist freely in space. A proton is an ion since it is the nucleus of the hydrogen atom. In our description of space storms we may use ions and protons almost interchangeably because they participate together in the complex processes of space, but in the near-Earth environment beyond the atmosphere, protons are usually the dominant ion.

The credo of charged particles might be 'May the force be with you'—it always is. Charged particles cannot escape the influence of electromagnetic force fields. Even though force fields can't actually be seen, physicists like to visualize them with imaginary lines in space and also with a direction associated with the lines. Lines closer together indicate stronger forces and the direction of the lines tells us how to compute the direction of the force.

In addition to familiar gravity, there are two force fields that may influence the motion of free charged particles in the vacuum of space. These are electric fields and magnetic fields. For our purposes, an essential distinction is that electric fields accelerate charged particles along the field. Positively charged particles are accelerated in the direction of the electric field and negatively charged particles in the opposite direction. Magnetic fields, on the other hand, can only accelerate moving charged particles in a direction perpendicular to the field (to the left for electrons and to the right for protons in an upward pointing field). In other words, electric fields can change the speed of a charged particle but magnetic fields can only change their direction. If things aren't confusing enough, put the two types of fields together and you get a different kind of motion altogether. This small poem may help clarify the matter—or it may confuse.

Lost in Space?

If ever you're lost
in the reaches of space,
a friendly charged particle
will find you your place.

In an upward field
of the magnetic type,
electrons turn left
and protons turn right.

In a westward field,
an electric beast,
protons head west
and electrons east.

Combine the two fields
and believe it or not,
electron or proton
toward Earth is your lot.

So follow an ion,
electron or two,
and I'm sure you'll know
just what to do.

John W. Freeman

More details about these and other exotic players in our story can be found in the Glossary and the 'Mathematical Appendix: A Closer Look', both of which can be found at the end of the book. Words explained in the Glossary are given in italic at their first occurrence in the text.

Vignettes of the Storm

Two hundred and fifty miles above the Earth, moving 7 kilometers a second, a gangly satellite with tubular sections and glistening wings is seen moving from the darkness of night into the blinding Sun (Figure 0.1, color section). As it crosses the terminator and moves into sunlight we recognize the International Space Station (ISS) with two tiny objects floating among the superstructure—astronauts on an extravehicular construction job.

The intercom to the EVA support crew crackles: 'you okay Greg?'

'Yeah. The glare from sunrise really blinds a guy for a few moments. Got to take a break to lower my sunvisor and acclimate. I'll get back to tensioning this strut in a second.'

'No problem.'

'Jake, the Sun botherin' you?'

'Negative—I'm in the shadow of the habitat.'

'Okay guys, I'd like to start our half-hour, suit-systems check list whenever you're ready.'

'Hardly seems necessary. Why interrupt the timeline?'

' We're overdue.'

'Okay—whenever.'

'O-2 pressure, Greg?'

'21.6.'

'Roger—21.6. Jake?'

'21.4.'

'Roger—21.4.'

'Coolant Temp? Greg?'

'23.8.'

'23.8. Okay. Radiation dosimeter reading, Greg? ... Greg, do you read me?'

'Yeah—this thing must be screwed up.'

'Why?'

'It's reading in the red zone—96 REMs and climbing!'

'Must be some mistake. Double check it when your eyes clear. Jake, how 'bout you?'

'Oh my God! Mine's in the red too!'

Suddenly the capcom from Houston Control breaks in: 'ISS—Houston, we have a report of an X class flare (Figure 0.2, color section) in the central meridian with solar energetic proton fluxes seen at L1. Terminate EVA immediately. Do you copy?'

'Roger—Houston. Terminating EVA. We've got high dosimeter readings here. Bummer! Crew, let's get these guys inside. Jake, Greg, head for the airlock. We're ready on this end.'

'On the way.'

'Oh no.'

'Jake, what's wrong?'

'I think my tether is hung up on the last strut I worked on. Can't reach the airlock with the slack I have. I have to double back and go around the other side. This will take a while. Can you see from the monitor where I'm caught?'

'No. Your tether is not in view.'

'Greg, I'm coming around to help.'

'No Jake. Get inside. I'll get this thing loose—Damnit.'

'Houston—ISS. How long do you think we have before it gets really hot out there?'

'ISS—Houston, SRAG reports a fast-rise, transient flare—about an hour 'til maximum proton flux hits Earth.'

Turning off his mike the capcom groans and curses to himself, 'How did this thing sneak up on us?'

From the Flight Controller comes the crisp message: 'All Consoles prepare for a rescue launch at mark plus three hours.'

In a room adjacent to Mission Control the Flight Director has already assembled a team of doctors with a teleconference line to radiation specialists. A routine construction EVA on the International Space Station has quickly turned into a nightmare.

*   *   *

It has been an uneventful evening in the forecast room of NOAA's Space Environment Center in Boulder, Colorado. The second shift is about to end and the crew is restless, ready to head home. Unnoticed by anyone, a thin blue trace on a video monitor chart takes a sharp swing upward. A few minutes later, a red trace lower down on the same monitor shoots upward and pegs flat at the top of the graph. These traces indicate a burst of X-rays and then an increase in the flux of high-energy protons from the Sun. They are plots of data from sensors aboard the GOES-8 spacecraft in geostationary orbit.

One of the forecasters just coming on duty starts her routine scan of the complex computer displays on the room-sized console. She notices the X-ray and proton plots and turns immediately to a nearby display of the solar disk from the SOHO spacecraft. There she sees that a giant solar flare has erupted from a region on the Sun directly facing the Earth. She whistles softly and says, 'oh-oh, we're in for a whopper.' She alerts the rest of the shift, then picks up the phone to call the Air Force 55th Space Weather Squadron. The NOAA space forecast crew begins a methodical assessment of the magnitude of the coming geomagnetic storm and begins distribution of alerts and warnings for NOAA's space-weather customers.

*   *   *

The SOHO spacecraft stationed 1,500,000 km from the Earth at L1, that delicate point in space where the gravity field of the Earth and Sun just balance, is routinely reporting conditions on the Sun and something called the solar wind. SOHO has transmitted the solar image used by the NOAA forecasters to verify the growing storm on the Sun. In addition to the bright flare, SOHO has also caught sight of a large arching prominence in the solar atmosphere high above the flare. The prominence expands until, now larger than the Sun, it forms a giant halo around the Sun. It gains speed as it explodes outward into space. Fifty hours later, the remnants of the prominence reach L1. SOHO dutifully reports a sudden jump in the speed of the solar wind to nearly 2 million miles per hour. Accompanying this increase in solar wind speed is a change in the direction of the magnetic field being dragged along by the solar wind. In less than an hour this cloud of high-speed ionized gas with its twisted magnetic field will reach the Earth and energize the Earth's magnetosphere.

*   *   *

The scene is a darkened communication satellite operations control room in Arlington, Virginia. Across one wall are monitors displaying live TV pictures. The first monitor carries the Winter Olympics ski jump competition. The others across the same row display a sitcom, a weather channel, and a Spanish language commentator. Only the sound from the Olympics monitor can be heard. The panel behind a long console desk has more monitors containing charts and graphs with technical data on the status of seven communication satellites. The night shift operations crew is gathered around the Olympics monitor.

Twenty-two thousand miles above the Earth the giant Comsat orbits with its antennas pointed toward the USA and Mexico. Suddenly, a small jet of gas squirts out from a miniature rocket thruster on the side of the spacecraft. Unlike the usual satellite orientation adjustment bursts which last only a few seconds, this time the attitude control thruster does not shut off. The satellite begins to tumble, slowly at first, then more rapidly.

Back in the satellite control room, all four TV monitors go blank and a warning beep is heard from the spacecraft status monitors. An ALERT warning indicator announces that spacecraft S-5 attitude control sensors report an out-of-limits condition followed by a loss-of-signal indication from the main transmitter. 'Oh-oh! Phantom commands again!', shouts someone as the operations crew scrambles to return to their stations. They prepare to send new commands to the out-of-control spacecraft. Several control console phones begin to ring almost simultaneously. 'Prepare to re-route the four video transponders from S-5 to S-7 on my mark,' says the lead controller into his headset. 'Damn, we'll lose the extreme eastern part of our antenna footprint. We've got no other choice!' The phones are ignored. A state-of-the-art, $250 million comsat has just been reduced to orbiting junk.

⋆   ⋆   ⋆

Somewhere in the Bahamas, a small sailboat is plowing its way eastward in seas that have been steadily growing rougher over the last hour. It occurs to the gray-bearded skipper that they are beginning to pick up the southern edge of the hurricane Elaine. He starts to fasten the safety line to his harness as he tightens his grip on the wheel. Without warning a crack is heard and the mainsail mast crashes across the starboard gunwale. The sail settles into the water and the boat begins to list dangerously. The skipper screams into the cabin for his wife to activate the Emergency Position Indication Rescue Beacon (EPIRB) that will send a MAYDAY call to an orbiting satellite. Unfortunately, the computers in Florida monitoring the EPIRB system will never receive the MAYDAY call. The southbound satellite overhead is no longer listening. Moments earlier it has been knocked from its proper orientation by strong magnetic currents associated with a nearby display of northern lights. The search and rescue will never take place.

*　*　*

Seven miles above Nova Scotia, American Airlines Flight 80 from Chicago's O'Hare airport is heading eastward through the night bound for Stockholm's Arlanda airport with 320 persons aboard. The pilot has just interrupted the in-flight movie to announce to passengers that there is an intense display of northern lights that can be seen by looking out the right side of the aircraft. After his mike is turned off, he comments to the first officer that he has never seen the aurora this far south before. A persistent, flashing orange light on the instrument panel interrupts the quiet conversation in the cockpit. The light indicates that the automatic system has lost lock on the radio beacon from the GPS navigation system. Instead of satellite navigation, the autopilot will now be relying on the onboard inertial navigation system. The pilot mumbles something to the first officer who sets about checking the onboard system.

By now, the cabin lights have been dimmed and a few excited passengers are straining to see the aurora borealis through the windows. Not far from the plane, they can see, dancing across the sky, a series of constantly changing rays of light that appears to diverge from far overhead and end abruptly at a bright lower border. The rays are mostly pale yellow and green with hints of red closer to the ground. They seem to form a curtain with many irregular folds that wiggle, disappear and then reappear elsewhere. In the distance other curtains can be seen flashing and shifting with folds that drift slowly westward. 'Oh that's beautiful! What makes the northern lights?' a small girl whispers to her father who is trying vainly to fall asleep.

*　*　*

'Capt'n O'Brien, sir, there's somethin' strange here in this last printout from the NORAD mast'r satellite log. It seems shorter than it oughtta be.'

'How so?'

'Well, we got 3,246 birds on the regular roster, not count'n' the launch of that Russian SPYSAT yesterday. Now the PO shows only 3,169 known birds tracked by the radars today.'

'Any pattern to the missing birds?'

'Lemmee check a minute. Yep, Holy Smoke! They're all LEO—polar orbits.'

'How come we didn't get a warning from Space Command?'

'Dunno, unless the link is down—heard somethin' bout a space storm.'

'How the hell can 75 satellites just disappear?'

'Wait, there's a new PO comin' in now. Damn, it shows a whole bunch of new birds! ...They've got some of the transponder IDs of the old guys. Looks like somethin' took the missing birds and shoved them into different orbits! The computer couldn't make the connection. What do you make of that?'

'I'll bet its got somethin' to do with that space storm and atmospheric drag. Let's send an email to 55th and see if they know anything.'

*   *   *

At a television station in San Francisco, a weather presenter begins preparation for his report on the 6 PM news. He switches on the monitor that provides his link to the GOES-8 weather satellite at geostationary orbit to check the current cloud cover conditions for the Bay Area, and to prepare some colorful graphics of hurricane Elaine threatening the east coast of the US. Instead of a beautiful false color, infrared image of the cloud cover over the South Atlantic, he finds a cryptic message on the screen, 'GOES-8 WEFAX temporarily out-of-service'. It does not occur to him that another type of storm, a storm in space, has disabled the weather satellite. Nor does he realize that this same storm in space is also interfering with data transmissions from another satellite; this one keeping watch on shear along the San Andreas fault line.

*   *   *

In the basement of the Pentagon, the night shift at the Defense Intelligence Agency Situation Center is preparing its routine morning briefing for the Joint Chiefs on overnight troop movements in Iraq. An aid enters with a report that BIGBIRD-1, the surveillance satellite positioned over southwestern Asia, has just begun to experience phantom commands, sudden and unexpected changes in operating modes, and loss of all digital photo-channels. The mood shifts abruptly to one of heightened concern as the group begins to consider whether or not to report this satellite problem as an indication of hostile activity, possibly from a ground-based microwave beam or laser.

*   *   *

In Montreal, Canada, a young family has just finished a late dinner and is heading for the living room for an evening of television. 'Dishes first!' shouts mom from the kitchen. No one seems to hear her. Instead, groans of frustration are heard from the living room. The TV cable channels do not seem to be working. A message on the screen indicates 'We are Experiencing Technical Difficulty with the Relay Satellite.' Suddenly, the lights go off. The entire city and most of Quebec Province are without power. It is nine hours before electric power is restored (Figure 0.3, color section). The cable TV channels remain out.

*   *   *

The production manger of a microelectronics plant near Torrance, California, is just turning into his driveway after the day shift. He notices, puzzled, that his garage door has opened before he keyed the transmitter on the sun visor. As the car rolls to a stop, his beeper sounds. A cell-phone call back to the plant tells him that the automated assembly line has begun to reject large numbers of microcircuits. He is aware of a growing migraine pain near the back of his head as he calls to his wife and backs out of the garage. Frustration is evident in her voice as she calls back after him that they have tickets for a play that starts in less that two hours.

*   *   *

Near the Ottawa River in Quebec Province, Canada, the Deep River Neutron Monitor Facility provides hourly measurements of the flux of neutrons produced by nuclear collisions between atoms of the Earth's atmosphere and galactic cosmic rays or energetic protons from the Sun. The record of such fluxes is usually one of steady, quiet, monotonous readings occasionally interrupted by slow decreases in the neutron flux as the stream of particles that enters the solar system from the Milky Way is temporarily blocked by a stronger solar wind. Occasionally, the neutron flux increases as streams of energetic protons from the Sun bombard the Earth.

About the time the Space Station astronauts are scrambling for safety, a graduate student on duty at Deep River awakes from having dozed off, at first dimly aware that an alarm bell is ringing. Shaking the cobwebs from his head, the student moves to the computer console and pushes a button to silence the alarm while checking the computer screen for diagnostics that will tell him the cause of the problem. The display shows that the neutron flux has risen to the top of the plot and 'pegged' the meter. Quickly checking the housekeeping data on the giant gas-filled cylindrical chamber that counts neutrons from space, the student determines that the monitor is operating correctly. The very high flux of neutrons is real! He has never seen the Earth bombarded by this many solar energetic protons before. He checks that the satellite data link to World Data Center A is open, initiates a data dump, and then reaches for the phone to call his advisor.

*   *   *

High overhead and to the southwest, a Strategic Air Command B-57 is cruising along the US–Canadian border over Lake Superior. Without warning, the huge aircraft banks sharply to starboard and begins a slow dive. The startled pilot quickly grabs the yoke and fights for control as the co-pilot cuts off the autopilot. The plane gradually returns to level flight. 'We have an autopilot memory fault indication' the co-pilot reports. 'I'll activate the backup system' and initiate a reboot. A short time later, the flight engineer reports over the intercom that the navigational computer has crashed. Avionics problems have disrupted a normally routine patrol mission.

These vignettes represent fictionalized, potentially real, events that could occur during giant storms in space.

How realistic are these scenes?

A major solar storm in 1991 deposited an intense flux of energetic protons in the vicinity of the Earth. These solar energetic particles caused significant irreversible damage to the solar power arrays on geostationary orbit satellites — shortening their useful life by many months. This same flux of protons would have produced a serious radiation health hazard for an unprotected astronaut if outside the Space Station for an extended time in a high-latitude orbit.

In January of 1994, an uninsured Canadian communication satellite valued at $225 million was lost during a storm in space. It is suspected that energetic electrons from space buried themselves in the surface of the spacecraft until a charge built up that resulted in an electrical arc. That arc caused a malfunction in the spacecraft attitude control system and the satellite spun out of control. Several other satellites experienced similar problems at about the same time.

In 1989, the Hydro Quebec power grid serving much of eastern Canada experienced a blackout that lasted nearly nine hours. Repair and losses from the blackout cost close to a billion dollars. The blackout was coincident with one of the largest space storms of the decade.

The study of space storms and the capability to forecast them has come to be called space weather. As our electronic systems become more vulnerable, space weather is having an increasing impact on our lives. In the following chapters we will explore the origins of space weather and its effects on human technological systems.

I   Two Kinds of Weather

Just as violent storms that rage in the Earth's lower atmosphere are manifestations of tropospheric weather, so also giant storms that take place in space surrounding the Earth are manifestations of what has come to be called 'space weather'. These space storms are only visible to humans through the dramatic northern and southern lights; however, their invisible effects are increasingly felt on our technological systems. There are many parallels between tropospheric weather and space weather.

It is an early Spring day in Houston. A strong damp wind is blowing from the south rushing to fill a trough of low pressure to the north. The weather forecast is for heavy storms as a front moves southeast across the area. Already the TV weather map shows a line of thunderstorms to the north. This brewing spring storm, like all storms, is the result of differences in atmospheric pressure, or pressure gradients as they are commonly called. We are familiar with dramatic barometric 'lows' of great storms like hurricanes and the ideal weather that comes from 'highs'. Driven by pressure gradients, great masses of air are pushed from the high-pressure regions towards the 'lows', creating winds.

Warm air is lighter than cold air and warm, wet air is even lighter. A hurricane is born when a large vertical column of warm, moist air fed by the warm tropical ocean rises upward like a giant chimney. This upward wind creates a low-pressure area at lower altitudes that becomes the eye of the storm. This low-pressure area causes air to rush horizontally inward from all directions in a vain attempt to balance the pressure. At the same time, a force caused by the rotation of the Earth, which turns straight-line motion into circular motion—the Coriolis force—causes the inward moving air to circulate counterclockwise in the Northern Hemisphere and clockwise in the Southern Hemisphere. This rotational velocity increases toward the eye, which only serves to increase the pressure difference and strengthen the storm.

If storms in the Earth's atmosphere are caused by winds driven by pressure differences, how can we talk about storms in space? There are certainly no winds of the type that generate hurricanes in space! We have been taught that space is empty. Well, space is *not quite* empty! What *is* there makes a world of difference.

Contrary to common belief, the space above the Earth's atmosphere, out to the Sun, and beyond, is populated by a tenuous gas made up of charged atomic particles, *electrons*, *protons*, and other *ions*. This space is also populated by magnetic and electric fields as well as a strange form of wave that can propagate only in the rarefied environment of space.

A gas consisting of charged (ionized) atomic particles is called a *plasma*. The plasma state is sometimes called the fourth state of matter after gases, liquids, and solids. While not common to earthlings, the plasma state is a common form of matter throughout the rest of the universe. The Sun is a plasma. Despite the extremely low density of the ionized gas or plasma of space, it still exhibits properties such as pressure, temperature, density and flow velocity that we associate with an ordinary gas. Just as we use terms like barometric pressure, air temperature and density and wind speed to describe atmospheric weather, we speak of the pressure, temperature, density, and velocity of the 'solar wind'. Further, just as the Earth's atmosphere exhibits global convection patterns such as the jet stream and lower-altitude circulation patterns like the trade winds, the plasma in space is subject to large-scale convection motion and winds. It seems appropriate then to think in terms of the 'weather in space'.

Our space weather begins with large pressure gradients that exist in the solar *corona*, the outer region of the Sun's atmosphere. The corona can be seen during an eclipse of the Sun as bright streamers extending outward several times the radius of the Sun. Figure 1.1 (color section) shows the solar corona during an eclipse. The corona has a temperature of several million degrees but cools as we move away from the Sun. Furthermore, there is a decrease in the density of the coronal gas with distance from the Sun. This means the pressure drops as we move away from the Sun creating a pressure difference, or gradient. This coronal pressure gradient is great enough to overcome the force of gravity and blow the coronal gas away from the Sun (see Figure 1.2, color section). This outward flowing ionized gas is known, again by analogy with terrestrial weather, as the solar wind. In one of the early triumphs of space science, Eugene Parker showed that the solar wind should become supersonic at a predictable distance from the Sun. Because the corona is a permanent feature of the solar atmosphere, the solar wind blows continuously away from the Sun to the orbit of Earth and beyond—all the way to the outer reaches of the solar system.

Even though we refer to it as a 'wind' the solar wind is a very tenuous plasma, having a density of only a few protons or electrons per cubic centimeter. It could not be felt if it were to blow against your face. On the other hand, its velocity greatly exceeds the velocity of any wind on Earth, being about 500 km per second or over a million miles per hour. Likewise, its temperature, as defined by the thermal motion of the atomic particles, is extremely high, in excess of 100,000 degrees.

In the vicinity of the Earth, the solar wind encounters the Earth's giant magnetic field. The solar wind is effectively blocked from penetrating the magnetic field because the magnetic force turns aside the solar wind protons and electrons.

Without the solar wind, the Earth's magnetic field would be a giant *dipole* spreading out in all directions with closed-loop field lines at lower latitudes and more radial lines toward the poles. Like any moving gas or plasma, the solar wind has pressure because of the thermal motion of the particles and also because of the flow motion (see Mathematical Appendix). The flow pressure of the solar wind is greatest in the direction of the motion, outward away from the Sun. The solar wind pressure confines the magnetic field of the Earth and compresses it on the sunward side and draws it away on the night side to form a long tail; an elongated magnetic bubble within the solar wind. This magnetic bubble, with the Earth at the center, is called the *magnetosphere* (see Figure 1.2, color section). The outer surface of the magnetosphere is the *magnetopause*. The region drawn out downwind is called the *magnetotail*.

The magnetosphere is the region surrounding the Earth where the Earth's magnetic field controls the behavior of charged particles in space and outside of which the solar wind sweeps past the Earth. The magnetosphere is invisible to human eyes but if we could visualize it the magnetosphere would appear as a giant bullet-shaped region, very much like a giant windsock pointed toward the Sun and extending about one sixth of the way to the Moon on the day side but with a long tail extending well past the orbit of the Moon on the night side.

The magnetosphere is the site of great storms in space. These storms, like their terrestrial counterparts, are driven by differences in pressure. How can pressure gradients exist in the magnetosphere?

Let's take another look at Figure 1.2. Does the flow of the solar wind over the surface of the magnetosphere have a familiar look? Perhaps you would agree that it looks very much like the wind flow past an automobile — better yet a convertible with the top down — like the upper drawing in Figure 1.3. The windshield forces the air up and over the car just as the magnetopause on the day side of the magnetosphere causes the solar wind to be deflected over (and around) the magnetosphere.

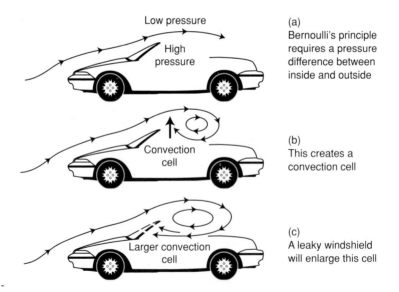

FIGURE 1.3 Illustration of the convection cell in an open convertible.

Have you ever had the pleasure of riding in an open convertible? Perhaps you noticed that, unexpectedly, the wind blows against the *back* of your head, or, if you have long hair, that your hair blows *forward or up* rather than back. Your hair may be blown in your eyes! (I highly recommend this experiment.) But why is this? Let's examine the situation further. As air streams over the top of the car it is moving fast relative to the air inside the car. In the early eighteenth century, Daniel Bernoulli developed a principle that states essentially: *where the velocity of a fluid is high, the pressure is low, and where the velocity is low, the pressure is high.* According to this principle the air streaming above the car must be at a lower pressure than the inside air. A chimney effect is created that pulls air upward. To compensate for this partial vacuum, air is pulled in from behind. This is also the reason a chimney will draft better on a windy day. A convection cell is created behind the windshield as shown in the center drawing in Figure 1.3. Pressure gradients drive the convection cell within the convertible.

Imagine further what would happen if I were to drill a series of tiny holes in the windshield—a horrible thought! If the windshield had small holes in it some air could leak out the front and join the slipstream. Then the convection cell would enlarge with forward moving air extending to the windshield itself as shown in the lower sketch in Figure 1.3.

What has all this to do with the magnetosphere? Our giant magnetosphere behaves like an open convertible! The 'stuff' of the magnetosphere that replaces the air in the convertible is the magnetic field of the Earth together with the dilute plasma that consists of magnetically trapped charged particles: protons, electrons, and ions. This 'stuff' constitutes the magnetospheric plasma. The solar wind plasma and the magnetospheric plasma are called *magnetohydrodynamic* (MHD) plasmas because they have embedded magnetic fields. These plasmas have pressure that is, in many ways, analogous to the air pressure of the Earth's atmosphere.

The magnetosphere exhibits pressure gradients and there are even convection cells (Figure 1.4). How can this be? The magnetic field of

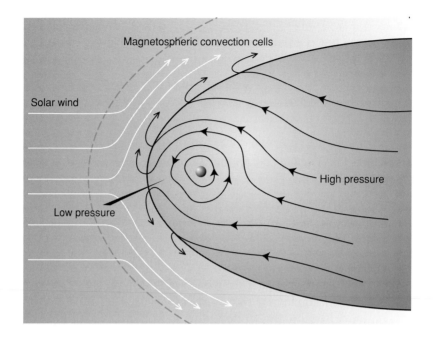

FIGURE 1.4 Illustration of the convection cells in the magnetosphere.

the Sun is dragged out from the corona by the solar wind as it speeds outward. This magnetic field, embedded in the solar wind as it encounters the magnetosphere, has twists and turns that are stretched remnants of the solar field. This field often has a direction opposite to that of the Earth at some locations on the dayside magnetopause. The intermixing of these contrary magnetic fields at the nose of the magnetosphere creates holes in the boundary between the magnetosphere and the solar wind so that magnetospheric plasma can leak out—much like the air through the holes we drilled in the windshield. This creates a low-pressure zone at the nose of the magnetosphere. Just as for low-pressure regions in the Earth's atmosphere, plasma from the magnetotail rushes forward in an attempt to balance the MHD low pressure in front. This results in a convection pattern that draws the low-energy magnetospheric plasma forward, around the Earth, and toward the Sun. The convection flow may turn away from the Sun at or just inside

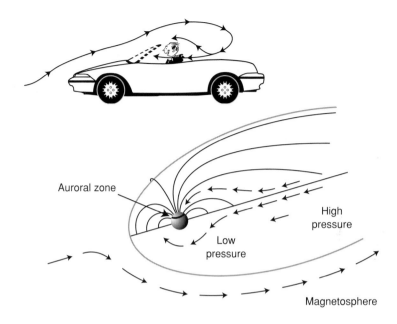

FIGURE 1.5 Comparison between the flow around a passenger in a convertible and the Earth as an obstacle to flow in the magnetosphere.

the magnetopause to form a convection cell similar to that in the convertible with the porous windshield (see Figure 1.4).

The analogy of the convertible/magnetosphere can be pushed still further. Our open car would not get far without a driver. What does the driver's head do to the convection cell? It acts as an obstacle to deflect the forward-moving air upward and also around. Our magnetosphere has an obstacle to forward motion as well! The obstacle in our magnetosphere is the Earth, or more precisely, its strong magnetic field close to the Earth. As illustrated in Figure 1.5, the Earth's magnetic field creates a high pressure zone in the center that forces the sunward moving plasma to move out around the sides and also upward along magnetic field lines into the auroral zone.

The plasma motion described here is motion of only the lowest energy or weakest component of the magnetospheric 'stuff', that component most easily pushed around by the MHD pressure gradients.

The more energetic particles lead a somewhat different life, which we will explore in a later chapter.

Just as the solar wind blows continuously, so the magnetospheric convection we have described goes on continuously. However, like the Earth's atmospheric wind, the solar wind is also variable. Changes in the solar wind result in changes in the convection pattern—in some cases the direction of flow may even be reversed. Most of the time, solar wind changes are gradual; however, just as atmospheric wind shifts accompany storms, radical changes in the solar wind speed, magnetic field strength and direction cause sudden and dramatic enhancements in the magnetospheric convection pattern. These dramatic changes in the convection rate lead to magnetospheric storms—storms in space. Changes in the solar wind that precipitate magnetospheric storms are caused by great ejections of matter from the Sun associated with violent outflowing coronal prominences called *coronal mass ejections* (or CMEs).

There is still one more analogy with the convertible. If our car could be accelerated to supersonic speeds, a shock wave would be formed out in front, like the shock wave that creates the sonic boom from a supersonic jet aircraft. Because the solar wind speed is faster than the speed of waves characteristic of the interplanetary medium, Alfvén waves, a thin *bow shock* wave must form ahead of the magnetopause. This shock wave serves to alert the onrushing solar wind that there is an obstacle ahead—to slow the solar wind to subsonic velocities so that it may flow gently around the magnetosphere. In the process much of the flow energy of the solar wind turns into heat increasing the temperature of the plasma flowing around the magnetosphere. We shall see that some of this heated plasma may ultimately find its way into the magnetosphere where it can play a key role in the coming storm.

2   The Saga of the Storm

In this chapter we follow the drama of the great storm associated with the events in the introductory vignettes. The reader is invited to try to identify the component of the storm that might have led to each of these events.

Our magnetospheric storm is born 150 million kilometers from Earth, in the atmosphere of the Sun. A loop of hot, ionized gas, many times larger than the Earth, arches high above the solar corona, grows slowly, then gains speed and blasts outward at a thousand miles a second (Figure 2.1). Expanding rapidly now, like a giant balloon filled with a million-degree gas, it is soon larger than the Sun and headed straight for Earth. We can see it coming. Digital cameras aboard a spacecraft orbiting between the Sun and Earth report a giant expanding halo surrounding the Sun (Figure 2.2, color section). We are watching a coronal mass ejection from the center of its bull's-eye.

As our coronal mass ejection or CME moves quickly outward it overtakes the slower solar wind ahead of it producing a violent shock wave at the leading edge (see Figure 2.3, color section). Protons trapped between the slower solar wind and the overtaking faster wind find themselves bounced helplessly back and forth and are quickly accelerated to high energies—like a ping pong ball trapped between a paddle and the table as the paddle is pushed down against the table.

Traveling at close to the speed of light, these high-energy protons soon escape from their magnetic bubble entrapment and thread their way along *interplanetary magnetic field* lines. Some find their way to the Earth well ahead of the CME. Their very high energy allows these solar energetic particles to penetrate the extended regions of the Earth's magnetic field and reach the top of the atmosphere over the north and south poles. Once in the upper atmosphere these solar energetic protons collide with the nuclei of atmospheric atoms and produce nuclear reactions resulting in a shower of secondary particles including neutrons. They also knock electrons from the atmospheric atoms and

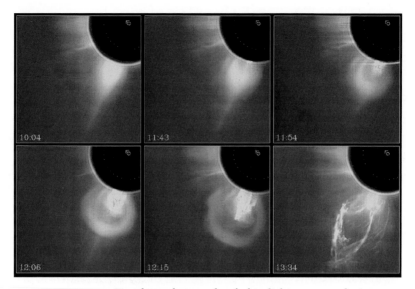

FIGURE 2.1 Time lapse photography of white light coronagraph pictures showing the evolution of a coronal mass ejection as it explodes from the Sun. The six images span a period of about three and a half hours. The CME is seen here in profile, from the side. Source: Coronagraph data from the Solar Maximum Mission, archived at the High Altitude Observatory. Primary image processing by J. Burkepile, High Altitude Observatory, National Center for Atmospheric Research (Boulder, CO).

molecules producing ions and free electrons, greatly increasing the number of electrons in the charged region of the Earth's upper atmosphere, the *ionosphere*. This enhanced ionosphere blocks and deflects radio waves at high latitudes causing a polar radio blackout. Radio signals at certain frequencies that normally have limited range can now propagate around the world.

This blizzard of energetic particles continues to bombard the polar region of the Earth for many hours, gradually diminishing in intensity until the flux falls below that of the steady drizzle of galactic cosmic rays that reaches the Earth continuously from outside the solar system.

A short time after the CME has left the Sun, a giant solar *flare* can be seen by cameras aboard spacecraft and by solar telescopes on Earth. The bright flare spews out an intense burst of X-rays, and radio noise.

The X-rays and radio noise travel toward Earth at the speed of light. The flare itself lasts for several hours before it can no longer be seen against the mottled solar *photosphere.*

Fifty hours later, the remnant of the CME reaches Earth in the form of a giant *magnetic cloud* carrying with it the twisted magnetic field drawn out from the Sun (Figure 2.3). The magnetic cloud, with its high-speed solar wind and helical magnetic field, engulfs and electrifies the Earth's magnetic field, initiating a great storm within the Earth's magnetosphere.

Increased pressure from the magnetic cloud compresses the magnetosphere so much that it shrinks to two thirds of its original size. This compression is promptly detected by magnetometers on the surface of the Earth, near the equator, as a sudden jump in the strength of the magnetic field.

But it is the twists and helical distortions of the interplanetary magnetic field (IMF) carried in the magnetic cloud that are devil's tools. During quieter times the IMF has a direction that lies mainly in the plane of the ecliptic, having been drawn out from the rotating Sun by the solar wind to form a giant Archimedean spiral, like the grooves in a phonograph record. In contrast, the twists in the magnetic field in the magnetic cloud take the form of a giant helical structure far larger than the magnetosphere itself. These twists give the interplanetary magnetic field strong components northward or southward out of the equatorial plane of the Sun and the Earth. It is the strong southward magnetic field component in the cloud that plays a major role in the excitation of the magnetosphere during our geomagnetic storm.

As this enhanced southward component of the IMF is swept rapidly against the Earth's magnetic field on the dayside of the magnetosphere, it finds itself pressed up against the Earth's magnetic field which points northward. A sensitive compass needle moved across the interface between the cloud and the Earth's magnetic field near the equator would change quickly from south to north. This configuration of two adjacent, oppositely directed magnetic fields is not tolerated for long. Instead, the two field regions quickly merge. In the process, magnetic energy disappears and then reappears as accelerated ions and electrons that have been sucked into the mêlée from both sides of the interface only to squirt out the sides at high speeds.

The merging of the solar wind's southward magnetic field and the Earth's northward magnetic field plays another vital role in our gathering storm. It removes magnetic field lines from the day side of the magnetosphere, throws them over the polar caps in the magnetotail and in the process lowers the 'pressure' in the day side. A stronger 'barometric low' is created. Magnetospheric plasma from the night side begins to move Sunward around the Earth at a faster rate to replace the missing flux. But the full response of the night side is temporarily, somehow, thwarted. Earthward flow from the long magnetotail is inhibited by counter-stresses building up and a great pressure imbalance develops.

Our long magnetotail swells and grows in importance as the solar wind removes magnetic field from the day side and flings it over the poles to the night side. On the northern and southern sides of the magnetotail we see long tubes of magnetic field that are the lines of magnetic field from the polar regions drawn back by the solar wind. Between these tail 'lobes', along the magnetic equator, lies a plane of dense plasma called the *plasmasheet*. Within the plasmasheet is a 'neutral' zone of very low magnetic field. At the center of this neutral zone the oppositely directed magnetic fields of the northern and southern lobes come into contact. A zone of denser plasma in the plasmasheet provides the particle pressure to keep the northern and southern magnetotail lobes at bay. Like our opposite magnetic field regions at the dayside magnetopause, the northern and southern lobes would like to devour one another.

Still the magnetotail has not provided its response to the increased Sunward convection already in progress in the inner magnetosphere. And so, the tension grows…

It is the plasmasheet that suddenly and violently provides the needed response. Without warning, as if propelled by a giant slingshot, plasma from the plasmasheet is suddenly flung Earthward to fill the void created by the increased convection. The northern and southern magnetic lobes merge with a violent release of energy. The result is a collapse of the tail near the Earth. Farther from the Earth, great bubbles of plasma emerge from the magnetic merging region and blow away from the Earth. Outer portions of the magnetotail may even be pinched off and blown away by the solar wind.

As the plasmasheet charged particles are pulled toward the Earth, they encounter the stronger, more dipole-like magnetic field of the inner magnetosphere (Figure 2.4). In so doing, they enter a regime in which different physical forces control their behavior. They no longer behave as a plasma or fluid but rather as individual particles, each of which moves according to the electric and magnetic forces it feels. The charged particles begin to move on individual paths as the electric and magnetic fields take control of their motion.

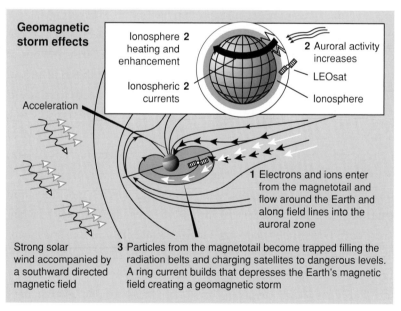

FIGURE 2.4 Diagram showing how an enhancement in the solar wind including a southward tilt to the interplanetary magnetic field results in a variety of changes that together constitute a geomagnetic (or magnetospheric) storm. The inset box at the top is a close-up of the Earth showing magnetospheric storm effects in the auroral zone.

Some of the electrons and ions pushing inward from the magneto-tail begin to move along the dipolar magnetic field lines and spiral down toward the Earth (Figure 2.4). Some penetrate all the way down to the atmosphere where they produce intense northern and southern lights displays. These auroral lights, seen only rarely at lower latitudes, are our only visible clue of the dramatic storm raging in space.

The downward flow of charged particles that produces the aurora also creates strong vertical electric currents and magnetic fields that interfere with control systems on low-altitude orbiting satellites.

Above the Earth's neutral atmosphere lies the ionosphere, a region where the atoms and molecules are highly ionized. The ionosphere is responsible for reflecting lower-frequency radio waves back to Earth and makes possible the long-distance transmission of AM radio stations and other long-wavelength radio communication. The iono-sphere is capable of conducting great electric currents.

Precipitation of the particles producing the aurora builds up charge in the auroral-zone ionosphere. To eliminate this charge, with the help of electric fields associated with the convection going on farther out in the magnetosphere, electric currents are produced in the ionosphere. These currents generate strong magnetic fields close to the ground that, in turn, induce similar currents in the Earth. These Earth currents may also pass through electric power lines. These storm-induced DC power line currents add to the AC power grid currents causing circuit overloads, tripping power grid circuit breakers or even exploding transformers.

As accelerated auroral particles enter the upper atmosphere and produce increased *ionization* another effect occurs: energy from the precipitating particles heats the atmosphere. The heated atmosphere expands upward and creates added atmospheric drag on low-altitude satellites, causing unexpected changes in their orbits.

The increased electron concentration in the ionosphere also changes the characteristics of radio propagation. Radio waves that normally pass through the ionosphere are now reflected back to Earth, signals from satellites are delayed slightly, and signals, which normally travel only a short distance, can suddenly travel thousands of miles.

Meanwhile, those electrons and ions rushing toward the Earth from the magnetotail that do not spiral down along magnetic field lines into the auroral zone are also up to mischief. As they approach the Earth they are turned sideways by the stronger magnetic field and forced to circle the Earth near the magnetic equator—electrons move eastward and protons and other positive ions move westward.

Electrons from this Earthward-rushing plasma easily reach the geostationary orbit where the orbit velocity of the satellite matches the rotation speed of the Earth and where many important satellites are parked. Satellites find themselves suddenly bathed in a gas of hot electrons. These electrons reach the surface of the satellites more frequently then their ion counterparts and charge the satellites to 5 kilovolts or more. Arcs begin to jump around on the spacecraft producing electrical transients that upset the spacecraft control circuits. To ground controllers it appears that phantoms have taken control of the spacecraft.

Other, more energetic electrons, protons and other ions pushing inward from the magnetotail are deflected around the Earth near the equator. Moving under the influence of the Earth's magnetic field, they form helical spirals along magnetic field lines bouncing repeatedly from one end of a field line to the other. At the same time, they also execute a sideways dance that causes them to drift in longitude around the Earth—electrons drift eastward and protons and heavier ions drift to the west. The longitudinal motion of the ions and electrons produces a giant ring of current that encircles the Earth. This 'ring current' causes the Earth's magnetic field inside the current to weaken in such a way that the magnetic field at the surface of the Earth is suddenly reduced. This drop in the Earth's magnetic field induces tiny voltages and currents in sensitive electronic circuits in use to test components at microchip factories or in long phone lines and transoceanic cables.

The reduction in the magnetic field Earthward of the ring current also causes the auroral zone to move toward the equator. The beautiful northern (or southern) lights usually seen only at high latitudes can now be seen at middle latitudes and sometimes even low latitudes.

Meanwhile some of the very energetic particles produced as the CME shock wave moves through the interplanetary medium have found their way deep into the magnetosphere. Here they can bombard satellites in geostationary and other high orbits. The high energy of these particles, mainly protons, allows them to pass through the protective cover glass on solar arrays that power these satellites. Once through the cover glass they penetrate the solar cells themselves and cause radiation damage that permanently reduces the efficiency of the solar array. The result is a shorter useful life for the spacecraft. Imaging cameras using charge-coupled devices (CCDs) may see streaks produced by the *solar energetic particles*.

Other solar energetic protons become trapped in the magnetic field, and, along with particles from the magnetotail, experience additional acceleration to form the *Van Allen radiation belts*. The radiation belts are two giant regions many times the size of the Earth in which electrons and protons with kinetic energies of millions of *electron volts* oscillate along the Earth's magnetic field lines bouncing repeatedly from one hemisphere to the other. These new radiation belts pose a different kind of hazard to spacecraft orbiting in the inner magnetosphere including the International Space Station. These magnetically trapped energetic particles may degrade the efficiency of the solar panels but, in addition, they bury themselves deep in the spacecraft skin. As they penetrate the insulating skin of the satellite, they build up charge, sometimes to the point where an arc path can be formed. A sudden, large discharge might disrupt or damage satellite electronics. Such an effect is referred to as 'deep dielectric discharge'. It may produce a larger, more destructive electrical transient than surface charging.

Solar energetic particles and higher energy galactic cosmic rays that come from outside the solar system will have sufficient energy to penetrate directly to sensitive electronic circuits within a satellite and upset individual circuits such as computer memory chips. Such events are called 'single-event upsets', or SEUs. It is believed that SEUs can permanently disable a $200 million spacecraft with a single miniature lightning stroke.

Astronauts aboard low-altitude spacecraft such as the space shuttle or the International Space Station might experience flashes of light in their eyeballs and receive increased doses of ionizing radiation from these energetic particles. Even more serious, unprotected astronauts performing outside construction work on the International Space Station or on an interplanetary mission to Mars could receive a heavy dose of radiation from solar energetic particles if precautions are not taken to provide protective shielding.

Solar energetic particle and galactic cosmic rays may affect avionics in high-altitude aircraft as well.

Twenty hours after the shock wave from the magnetic cloud hits the Earth, the magnetic field in the cloud has slowly rotated to a harmless northerly orientation and the wind speed has dropped to a gentle 350 kilometers per second. From the outside it appears that the storm has passed. Within the magnetosphere, however, the energetic electrons in the Van Allen belt are still climbing. The slow buildup of their deadly flux has not passed. The great distortions of the Earth's magnetic field are only now beginning to subside.

But, for earthlings, the storm has passed. The aurora has returned to its normal nightly parade of quiet arcs. Power blackout alerts are over. All is quiet on the Sun—for now.

3   Weather Stations in Space

The answer, my friend, is blowin' in the wind

Blowin' in the Wind, Bob Dylan

Like a lonely sentinel standing on a hill, the Advanced Composition Explorer (*ACE*) spacecraft hovers suspended between the gravitational pull of the Sun and Earth. From this vantage point upwind, ACE can warn of changes in the solar wind that herald a *magnetospheric storm* headed toward Earth.

ACE is 'parked' at the First Lagrangian point, *L1*, a point in space where the gravitational potentials of Earth and the Sun form a small gravitational potential hill 1.5 million kilometers from Earth and 150 million kilometers from the Sun (Figure 3.1, color section). At this point only a small amount of fuel is required to keep the spacecraft on a line between the Sun and the Earth. From this upstream point ACE can continuously monitor the solar wind headed toward Earth.

Like its terrestrial weather station cousins, ACE is equipped with instruments that provide information on the speed, direction, temperature and even density of the solar wind. Together, these measurements permit the determination of the dynamic pressure of the solar wind, the equivalent of our terrestrial barometric pressure. The dynamic solar wind pressure determines the extent to which the magnetosphere will be compressed by the intensified solar wind.

The ACE space weather station measures an additional physical quantity, the magnetic field carried along toward the Earth embedded in the solar wind. The interplanetary magnetic field, or IMF, plays the lead role in creating violent magnetospheric storms. As we have already seen, it is the merging of magnetic field lines from the solar wind with those of the Earth that produces the drop in pressure on the day side that drives increased magnetospheric convection and the resulting magnetospheric storms. The IMF is a *vector* and ACE measures each of its three components. The components of greatest importance to the generation of magnetic storms are those that can interact with the Earth's magnetic field near noon at the equator and at higher latitudes. These are the north–south component and, to a lesser extent, the east–west component of the IMF. Because the Earth's magnetic field points mainly northward near the equator on the day side of the magnetosphere, and because the greatest merging can occur when the fields are opposite to one another, a strong, continuous southward IMF produces great geomagnetic storms. In addition, a higher wind speed carries the magnetic field toward the Earth more rapidly, enhancing the effect of the southward IMF and strengthening the solar wind disturbance on the magnetosphere.

Magnetic clouds are solar wind disturbances that carry prolonged periods of southward IMF. They are large regions in the solar wind which carry a strong interplanetary magnetic field that rotates slowly as the cloud passes the Earth. As we have already seen, magnetic clouds are believed to be the remnants of coronal mass ejections that have been blown out from the Sun by the force of the solar wind, much like an explosion carries debris. As a giant magnetic cloud passes the Earth, the IMF can remain pointed toward the south for the better part of a day. This southward IMF interacts with the Earth's magnetic field to generate a giant magnetospheric storm.

In addition to the enhanced convection produced by the southward field in the magnetic cloud, the cloud itself may be preceded by an interplanetary shock wave that shakes up the Earth's magnetic field. As this shock compresses the magnetosphere it generates shock waves within the magnetosphere producing a variety of types of MHD and electromagnetic waves. The magnetosphere rings like a giant gong. The Earth receives a giant wake-up call that a major space storm is upon us. As the shock travels across the magnetosphere it induces an electric impulse that can accelerate particles trapped in the Van Allen belts to higher, more dangerous energies.

We can see now the importance of ACE's location upwind of the Earth. From here it can warn us of impending trouble. A magnetic cloud passing ACE, traveling at a speed of about 600 kilometers per second takes about 40 minutes to reach the magnetosphere. On the other hand, the data on solar wind conditions captured by the ACE instruments are radioed to Earth with the speed of light and reach our space weather forecast centers in 5 seconds, well ahead of the solar wind disturbance itself. This highly valuable lead-time gives forecasters time to run models and make an assessment of the effects that can be expected from the disturbance headed our way.

The role of ACE is indeed that of a storm warning station in space; but ACE is not alone in this task. Other spacecraft at the L1 site keep watch on the Sun itself using digital cameras (imagers), which scan the Sun in several wavelengths. The *SOHO* and YOHKOH spacecraft can see changes in the Sun's corona that are precursors of flares and CMEs. Coronal mass ejections directed toward the Earth are not easily detectable until they become larger than the Sun itself and appear as a halo surrounding the Sun. However, the tremendous forces in the corona associated with the launch of these eruptions generate waves on the Sun that may be used to pinpoint the site of the CME. Images of the Sun obtained from these spacecraft alert forecasters to the possibility of trouble on the way.

Closer to home, within the magnetosphere itself, Earth-orbiting satellites serving as space weather stations are also watching for signs of storms in space. A variety of satellites orbiting the Earth are monitoring additional space weather indicators.

The *geostationary* orbit is a particularly useful orbit. At an altitude of almost 36,000 kilometers, the orbit velocity of a satellite in a circular orbit near the Earth's equator exactly matches the rotation of the Earth. A satellite in this orbit appears to hover above one spot on the Earth. This is the orbit of choice for most communication and weather satellites. At the geostationary orbit, the *GOES* satellites, which are the *NOAA* weather satellites that provide cloud images of the Earth, also carry important space weather monitors. These include detectors that measure the X-ray *flux* from the Sun (Figure 3.2a, color section). Major solar flares produce large fluxes of low-energy X-rays. Near the Sun, these X-rays would be lethal but by the time they reach the distance of the orbit of Earth their intensity has been reduced many times over, to a level less that a watt per square meter: a thousandth of the energy flux from sunlight. Nevertheless, these X-rays serve as a good indicator of the start and of the magnitude of a solar flare. Because they travel at the speed of light, the X-rays arrive at Earth in about 8 minutes, before any of the energetic particles and long before the solar wind disturbances. Along with the solar images, the X-rays give forecasters an early warning of the on-coming storm.

The GOES satellites also carry energetic electron and proton detectors (Figure 3.2b, color section). These proton detectors herald the arrival of energetic protons spewed out by the flare or accelerated by the shock front that moves ahead of the CME. These protons arrive at Earth well ahead of the solar wind disturbances. The fluxes of these protons, as seen by the geostationary satellites, rise abruptly an hour or less after the flare or after the CME leaves the Sun. Like the X-rays, the flux of energetic protons peaks and then decays away slowly. Subsequent peaks may signal the occurrence of repetitive outbursts on the Sun.

The energy of these particles from the Sun ranges from a few million electron volts (MeV) to over a billion electron volts (BeV). These are the particles that damage the solar power systems on satellites and can be harmful to unprotected astronauts. The ACE spacecraft and the GOES geostationary orbiting satellites both tell us of the approach and presence of these solar energetic particles. However, because they travel at close to the speed of light, there is very little time to take protective action.

The magnetic field of the Earth deflects these energetic particles away from the equatorial region, but at higher latitudes, near the poles, they can reach the top of the atmosphere where they increase the electron content of the ionosphere and produce nuclear reactions that let ground-based detectors know of their presence. *Neutron* monitors and instruments that measure the electron content of the ionosphere (*riometers*) signal the penetration of the solar energetic protons into the atmosphere. These as well as other instruments on the ground also play important roles as space weather stations.

A little while later, as the magnetic cloud hits the magnetosphere, the GOES satellites also provide data on the development of the magnetospheric storm by reporting changes in the fluxes of Van Allen belt particles, electrons, protons and ions with energy much lower than the energetic protons from the Sun. Here the evidence gets more complicated because the signature of the storm as seen in the lower-energy trapped particles depends heavily on where the satellite is located at the onset of the storm. For a satellite on the day side of the Earth the first clue to a major storm is a sudden drop in the flux of trapped particles at all energies as they are swept out of the magnetos-phere across the magnetopause. On the other hand, a geostationary satellite near midnight will see a jump in the flux, particularly for the lowest-energy electrons, as these electrons are driven into the inner magnetosphere from the magnetotail. As the satellite moves toward the dawn meridian, the energy of the electrons grows as the flux continues to increase. The more energetic particles slowly build in intensity until the magnetosphere is loaded with a fresh supply of trapped particles.

As these new particles drift around the Earth they produce the ring current. This current has the shape of a giant ring of current flowing from east to west around the Earth. The ring current in turn generates a magnetic field that can be felt all the way down to the surface of the Earth.

On the ground, other terrestrial space weather stations called *magnetometers* swing into action and transmit the detection of the ring current by relay satellite to scientists around the world. Data from these ground magnetometers are used to construct several 'indices' that provide a measure of the strength of the storm going on in space. One index, called *Dst* for 'Disturbance storm', provides a direct indica-tion of the ring current. Dst is a measure of the difference between the normal quiet-time terrestrial magnetic field and the changes produced by the storm mainly due to the particles encircling the Earth. Because

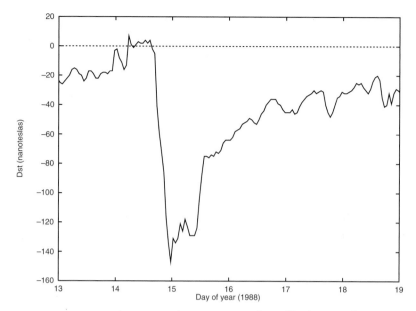

FIGURE 3.3 A magnetospheric storm as indicated by the Dst index.
The dotted line near the top of the figure represents a quiet, prestorm
condition. The great dip in the curve represents the 'main phase' of the
storm when conditions in the magnetosphere are the most disturbed.
The Dst index is created from magnetic data taken from ground-based
magnetometers around the world. Source: Rice University Department
of Physics and Astronomy.

the ring current lowers the magnetic field of the Earth near the equator,
during the main part of the storm when the ring current is strong, Dst is
negative (Figure 3.3). In the early part of the storm, when the solar wind
pressure is compressing the magnetosphere, Dst is slightly positive.
Dst is used extensively as an indicator of the strength and duration of
magnetospheric storms.

A second index, called the *Kp* index, provides a more general
measure of storm activity since it is sensitive to higher latitude
magnetic disturbances as well as low latitude activity. Kp has been
used for many decades by space scientists to quantify the intensity of
magnetospheric storms. More recently, scientists working for the US
Air Force have developed a version of the Kp index using fewer ground

magnetic observatories. Use of fewer ground magnetometers allows this 'pseudo' Kp to be generated in near real-time for use in space weather nowcasting. More recent technology uses an artificial intelligence code with *neural networks* trained to forecast a pseudo Kp from solar wind data transmitted from the ACE spacecraft. This gives forecasters at NOAA's Space Environment Center in Boulder, Colorado, a measure of the intensity of the approaching storm even before the magnetic cloud has hit the magnetosphere.

Other ground observatories also play a role as space weather stations. Magnetic observatories in the auroral zone provide data for the *AE* index, a measure of currents that flow in the auroral zone ionosphere. Photometers can see and measure the light from intense auroral activity and provide another measure of the storm raging in space. Powerful radars, trained to look at the polar ionosphere, show us how the motion of charged particles in the upper atmosphere is responding to electric fields impressed on the magnetosphere by the solar wind. These mirror the giant magnetospheric convection patterns described in Chapter 1.

Back in space, satellite-borne imagers pointed toward Earth give us a large-scale look at the development and breadth of the aurora. Another powerful tool is being developed. Space-based imagers capable of remotely sensing activity in the magnetosphere itself also allow us a look at the development and intensity of each phase of the storm. The IMAGE satellite has a unique set of remote sensors able to trace the flow on magnetospheric plasma. IMAGE represents a new era in space weather stations.

4 Lights in the Night: The Signature of the Storm

We would watch and watch the silver dance of the mystic Northern
   Lights.
And soft they danced from the Polar sky and swept in Primrose haze;
And swift they pranced with silver feet, and pierced with a blinding blaze.
They danced a cotillion in the sky; they were rose and silver and shod;
It was not good for the eyes of man — 'twas a sight for the eyes of God.

Robert Service, *'The Ballad of the Northern Lights'*

It is a beautiful fall evening on the south shore of Lake Superior.
Despite the cool offshore breeze, the sand beneath us feels warm as we
lay on the beach and watch the steady arc of a satellite moving slowly
against the background of stars. The crescent moon has already set and
the night is dark and clear. Behind us the Milky Way stretches brilliantly
across the sky. Low in the north, the Big Dipper points to the
pole star. Beyond Cassiopeia to the east we can see the large open
constellation Andromeda. As we strain to catch a glimpse of the faint
Andromeda Galaxy, we suddenly become aware of dim, slowly
changing forms far to the north. They first appear as giant, pale yellow-
green draperies or rays hung from space, slowly changing shape,
appearing and then disappearing. Gradually, as we watch, we notice
that the rays begin to move toward us. The bottom of the rays, where the
light is most intense, forms quiet arcs that drift slowly across the sky
(Figure 4.1, color section).

The conversation continues as the night wears on. After midnight
someone notices an increase in the lights in the sky to the north. The
quiet arcs have begun to break up into multiple arcs or pieces of arcs.
Slowly these arcs drift toward us and the motion becomes more
violent; the colors more intense. Reds can now be seen occasionally.
Almost before we realize what has happened, the action is directly
overhead and we are looking up into a vast array of lights in the form of
curtains and rays that converge toward the zenith. We are looking
upward into an auroral corona. The lights are changing and dancing

continuously now with waves and flickers that flash across the sky. The colors are intense and yet subtle.

Over the next hour the motion slowly subsides and we now see single isolated beams or patches of light. Gradually these subdued forms retreat once again to the north where they become quiet arcs like those seen earlier in the evening. These arcs become more difficult to see as a faint hint of dawn grows in the eastern sky.

We have witnessed the auroral display associated with an intense storm in space. Such a display is called an auroral *substorm*.

We are not the only ones this night interested in the display of northern lights. The astronauts aboard the space shuttle Discovery have watched the same auroral substorm from space (Figure 4.2, color section).

Yet another pair of 'eyes', the ultraviolet imager and the Visible Imaging System (VIS) imager aboard the POLAR spacecraft, high above the polar cap, have been keeping tabs on our northern lights ministorm (Figure 4.3, color section). These eyes have certain advantages over us, and the astronauts. They can see the whole polar cap at a single glance, and they can see beyond the visible into the ultraviolet (UV) wavelengths. When seen in UV, the auroral arcs extend around to the dayside and encircle the polar cap in what is called the auroral oval.

Spaceborne cameras such as those on POLAR have helped scientists develop our understanding of the substorm. However, studies of the auroral substorm began with the pioneering work of Kristian Birkeland near the turn of the twentieth century. These studies were followed in more recent times, before the era of satellites, by Syun-Ichi Akasofu. Using ground-based cameras that could photograph the entire sky from horizon to horizon, Akasofu was able to discern the various phases in a typical auroral substorm. He found that, before the substorm, the aurora is characterized by long, featureless, quiet arcs that cover most of the night side of the Earth. The substorm development itself consists of an intensification, usually near the midnight meridian, that grows as a bulge toward the pole. At the same time, the southern edge of the auroral zone moves toward the equator. As the active region near midnight grows, it may also develop a strong surge that propagates away from midnight toward the west. Meanwhile the disturbance expands eastward toward dawn as well. This phase of intense activity is referred to as breakup. Breakup often starts near midnight and then spreads eastward and westward until the entire night side of the auroral zone is the scene of bright, highly disturbed auroral arcs or rays.

An observer on the ground sees this scene pass over him as he rotates with the Earth under the pattern of the auroral oval. As close as it may seem, the great pattern of the aurora is fixed in the distant magnetosphere. This pattern is traced on the upper atmosphere by electrons sliding down the long lines of magnetic force that reach to the night-side magnetosphere near the equator. Most of the auroral light is produced at an altitude of about 100 kilometers where the atmosphere is thin and the electron ionization lasts long enough to produce the glow we call the northern lights. It is as if an electron gun in some giant television set were painting a picture on the sky from above. The yellow-green auroral glow results from the excitation of oxygen atoms in the atmosphere just as our television picture emerges from the excitation of phosphors on the screen. The large-scale pattern itself is fixed in the reference frame of midnight because the solar wind holds the magnetosphere tightly in its grip with the magnetotail pointing forever away from the Sun. The auroral oval is a Sun-oriented pattern.

A fortunate observer may witness more than one substorm in a single night. Substorms repeat themselves with regularity. The frequency and strength of substorms depends on the intensity of the larger magnetic storm in progress farther out in the magnetosphere. A typical substorm lasts only an hour or so; however, during a very strong magnetospheric storm auroral substorms will appear more closely spaced and in rapid-fire succession.

The auroral substorm is the only manifestation of the substorm directly visible to man. Another more harmful manifestation of a substorm is rapid changes in the magnetic field near the auroral zone due to surges of ionospheric current associated with the aurora. These 'auroral electrojet' currents, as they are called, produce deep positive or negative excursions in the nearby magnetic field. These sudden changes in the magnetic field may induce currents in long power lines resulting in overloads on electric power grids. The overloads may cause power blackouts, a much less desirable manifestation of the substorm than our beautiful northern and southern lights.

These magnetic disturbances near the auroral zone are called polar magnetic substorms. Like the auroral substorms they may last an hour or so and are another component of the substorm. As with the geomagnetic storm itself, magnetic indices have been devised to measure the intensity of a polar magnetic substorm. One aspect of space weather forecasting is focused on predicting these indices and even the local magnetic variations caused by auroral electrojets near power transmission lines and transformers.

Like ocean currents that converge to form the Gulf Stream, electric currents feed the horizontal ionospheric currents of the electrojet from regions outside the auroral zone. Some of these currents flow toward (or away from) the Earth along magnetic field lines from far out in the magnetosphere and are the result of distant electric fields.

The behavior of the visible aurora and ionospheric currents during a substorm reflect a number of complex processes taking place far out in space. If we could step out into space and watch these processes with eyes that can see magnetic and electric fields, we might see something like this: a hundred thousand kilometers out in space the interplanetary magnetic field embedded in the solar wind makes connection with the magnetic field of the Earth as it sweeps past. The circular auroral zone surrounding the pole bounds the region of field lines that connect to the solar wind field. These so-called 'open' field lines, emanating from the polar region, get tugged backward over the poles as they are pulled along by the solar wind. Like a rotor in some giant generator, this motion of the magnetic field lines over the poles and into the magnetotail induces an enormous electric field inside the magnetosphere. It is as if a giant battery has been connected to the magnetosphere with the positive terminal on the dawn side and the negative terminal on the dusk side. An electric field stretches across the entire magnetosphere pointing from dawn to dusk. This electric field influences the motion of charged particles everywhere in the magnetosphere. As we have seen, one effect is the Earthward motion of electrons and ions inward from the magnetotail. Electrons and ions part company as they approach the inner magnetosphere to form the ring current, the hallmark of a magnetospheric storm. Near the equator the same large-scale electric field drives a current in a sheet across the midplane of the magnetotail. This current creates its own magnetic field that stretches the Earth's dipole field on the night side to form the long magnetotail.

Magnetic field lines that arch upward from the auroral zones pass through the tail and through this magnetotail current sheet. Suddenly, as if triggered by a giant switch, part of the sheet current is deflected from its path across the tail down along magnetic field lines and into the auroral zone in the midnight longitudes. Currents surge from east to west across the ionosphere just as violence erupts in the midnight auroral display. The current sheet electric field has been temporarily short-circuited through the aurora. Electrons can move unimpeded along the magnetic field and are accelerated to tens of *kiloelectron volts* by the electric field. As we have already seen, the colors of the aurora are produced by these energized electrons plunging into the Earth's upper atmosphere and exciting atoms of oxygen and molecules of nitrogen.

We see now how our dazzling display of auroral lights is connected to the motion of the solar wind past the Earth many tens of thousands of kilometers out in space. Indeed, the electrons that form our dancing auroral beams may have begun their journey from the Sun to our night sky only several tens of hours ago. We can also see that the auroral substorm and polar magnetic substorm are the byproduct of the large-scale magnetospheric storm. While we understand this general picture, there are still a number of unsolved puzzles related to the aurora. Despite an enormous amount of research, we don't yet understand the episodic nature of substorms and how they are triggered. In the next chapter we will explore in greater detail the mysteries of the magnetospheric storm itself.

5 A Walking Tour of the Magnetosphere

The real voyage of discovery lies not in seeking new
landscapes ... but in having new eyes.

Marcel Proust, 'The Captive' (*Remembrance of Things Past*)

Imagine for a moment that we can leave the comfortable cocoon of our atmosphere near the surface of the Earth and stroll through the near vacuum of space into the magnetosphere. As we journey outward, several thousand kilometers above the surface of the Earth we find ourselves bathed in a swarm of electrons and protons spiraling madly about magnetic field lines as charged particles like to do (tour stop 1, Figure 5.1, color section). They race around and around in great circles. The faster (more energetic) particles move in larger circles. The electrons spiral one way and the protons spiral around the lines of magnetic force in the opposite direction.

As we try to follow the path of a single electron, we discover that it is not just moving in a circle but also streaming northward (or southward) along the magnetic field, and ultimately downward toward the Earth. As it approaches the Earth, it moves in a tighter and flatter spiral. At a certain altitude it reverses its motion along the magnetic field line and heads back up towards us, then past us and downward into the opposite hemisphere. At some point, it 'bounces' again and heads back toward us once more. We are witnessing the spiral and latitude bounce motion of a geomagnetically trapped particle (Figure 5.1).

These 'trapped' particle orbits were first investigated by Carl Störmer in the early part of the twentieth century. Discovery of their population with real particles was the first great scientific discovery of the space age. James A. Van Allen and his co-workers at the University of Iowa found hints of the trapped radiation belts with the first US satellite Explorer I in 1958. The intensity of these trapped particles was so high that the Geiger counters on Explorer I saturated. Later satellites with more sophisticated detectors and telemetry were required to confirm the discovery.

How can we understand this bizarre particle motion? It defies everyday experience because it is possible only in the near vacuum of space. Furthermore, electrons and protons are atomic particles far too small to be seen by eye. Fear not, our walking tour will lead us to understanding.

Electrons are negatively charged particles. All charged particles experience a force when moving with any component of their velocity perpendicular to a magnetic field. Imagine yourself at the equator riding along on an electron speeding upward, directly away from the Earth. All the while, you are moving through the ever-present northward pointing magnetic field of the Earth. You will feel the electron continually drawn inexplicably toward the left. After a short time the electron will have turned around and you are headed toward the Earth. Then later you find yourself headed back out again—around and around you go in endless circles. The electron is being forced into circular motion by the magnetic force. This force is proportional to the electron's velocity perpendicular to the magnetic field and to the intensity of the magnetic field. (A positively charged particle, such as a proton, will feel a force to the right.) The magnetic force does nothing to change the speed of the electron (or kinetic energy), only its direction.

Now let's hop aboard a new electron. This time our electron, in addition to moving in a flat circle, has some velocity in the direction of the magnetic field. Our path is now a helix centered on the magnetic field line. The pitch of this helical path relative to the field line is called the 'pitch angle'. The helix is rapidly carrying us away from the equator. Because we are in the Earth's dipole magnetic field, as we move along the field line, two things happen: first, the field line starts to slowly curve downward toward the Earth; and second, the intensity (or strength) of the magnetic field begins to increase. As the field curves downward, so also our helix curves downward like a bent corkscrew. As the magnetic field intensifies because we are getting closer to the Earth, the circular force on our electron becomes stronger and the electron moves in ever increasingly smaller circles. The helix curves Earthward and tightens.

Now something still more remarkable happens. Because the magnetic field must curve downward to meet the Earth and because the magnetic force on the electron is at all times perpendicular to its motion, there is a small component of that force backward along the field line. This force reduces the pitch angle of the electron until at some point the helical motion vanishes and for an instant the electron is in a flat circle just like our original electron back at the equator. But the backward magnetic force does not turn off at the flat circle. Instead, it causes the electron to start helical motion back up the magnetic field line. The electron has reached its 'mirror point' and is being reflected away from the Earth (Figure 5.1). The helix once again stretches out and opens up. Upon crossing the magnetic equator, the field line starts to plunge toward the Earth again, and the process repeats itself, this time in the southern hemisphere.

While what we have witnessed here in our wild ride seems unfathomable, there are even more surprises in store. If we continue our ride on the same electron for a while longer, we notice an amazing thing. In addition to its dizzy helical bounce motion from hemisphere to hemisphere, our electron is also slowly drifting eastward, around the Earth. Apparently additional forces are at work. Looking more closely we notice that the arc our electron makes when projected onto the equator is not a perfect circle as we had originally thought. In fact, the electron makes a tighter circle on the side of its trajectory closest to the Earth and a larger circle on the side farther from the Earth. The effect of this is that each apparent circle carries the electron slightly eastward accounting for the electron's eastward drift around the Earth. This is called 'gradient drift' because it arises from the decrease in the intensity of the magnetic field with increasing distance from the Earth. A similar effect, 'curvature drift', also eastward, results from the dipolar curvature of the magnetic field lines. The electron's eastward drift due to the gradient and curvature forces produces a current in the westward direction in the shape of a giant loop around the Earth, like a hula-hoop. If we leave our electron and hop aboard a proton, we shall see that it executes all the same types of helical motion and gradient and curvature drift only in the reverse direction and with larger circles. The proton drifts westward but since it is positively charged compared with the electron's negative charge, the combination of the two yields an even stronger ring of current around the Earth.

Like any electrical current, this 'ring current' produces its own magnetic field. As the number of trapped particles builds up through the introduction of new particles from the magnetotail, the ring current magnetic field grows in intensity and begins to compete with the Earth's field. As we have seen, magnetometers at the surface of the Earth, particularly near the equator, measure the reduction in the Earth's magnetic field produced by the ring current. The magnetic storm index, Dst, is a measure of this reduction of the Earth's field due to the intensity of geomagnetically trapped electrons and protons. In this way Dst indicates the strength of the magnetospheric storm.

Let's now leave the region of the inner magnetosphere and travel outward on the night side, toward the magnetotail. As we approach the geostationary orbit we notice that the magnetic field is becoming stretched away from the Earth to become more elongated (tour stop 2, Figure 5.1, color section). A bit farther still we see swarms of lower energy electrons and protons advancing inward from the magnetotail.

At the same time we become aware of the presence of an electric field that seems to stretch from east to west across the great expanse of the magnetotail. It is this electric field that seems to be forcing the electrons and ions from the tail inward toward the geostationary orbit. We notice that the drift of the electrons and ions is at right angles to both this electric field and the magnetic field.

To investigate this strange motion we focus in on a single proton circling a magnetic field line. We know that, unlike a magnetic field, an electric field can accelerate a charged particle in the direction of the field. This case is more complicated, however, because of the simultaneous presence of a magnetic field. As the proton moves in the direction of the electric field it gains speed. This increases the radius of its circular motion about the magnetic field so it swings in a wider circle. Soon the proton has swung completely around and is moving opposite to the electric field. At this point the proton slows down causing it to move in a tighter circle for the next half of the orbit. The result of this succession of wider and tighter half-circles strung together is a cycloidal motion that results in a drift toward the Earth. You have probably seen motion like this before. Viewed from above, the trajectory of the proton looks very much like the prolate cycloid formed by the reflector on a bicycle wheel seen crossing headlight beams at night. The reflector goes round and round with the wheel but at the same time moves forward with the bicycle producing a sort of jerking forward motion.

Let's zoom in on an electron. Here we see a similar effect. Even though the direction of circular motion about the magnetic field is reversed, the electron has drift motion in the same direction as the proton. Both particles move inward toward the Earth.

As we watch the plasma sweep inward from the magnetotail, toward the stronger magnetic field, we notice that protons begin to drift toward the west and electrons toward the east around the Earth. Gradient drift begins to take control of the drift direction. These new particles add their contribution to the ring current.

Catching our breath a moment at tour stop 2, we notice that the inward sweep of tail particles is not steady but abruptly intensifies and then slowly dies away. Each new intensification leads to a fresh injection of particles into the inner magnetosphere and is accompanied by sudden changes in the magnetic and electric fields. The electric field suddenly weakens and the magnetic field strengthens and becomes more dipole-like.

Could there be a connection between these periodic injections of charged particles from the magnetotail and our auroral substorms? The answer is not long in coming.

Before leaving stop 2 we also notice that some of our new particles from the tail are being swept northward and southward along magnetic field lines, away from the equator, toward higher latitudes. An invisible force is drawing them down into the auroral zone around each pole. That force is the electric field that was creating the inward drift from the magnetotail. This electric field now extends along the magnetic field lines arcing downward toward the Earth. It can accelerate charged particles into the Earth's atmosphere to create the marvelous auroral lights we saw during our night on the beach.

As our electrons plunge into the atmosphere from above we can see that they have gained sufficient kinetic energy to knock atomic electrons from the atoms of the atmosphere and excite these atoms so they emit light. The color (wavelength) of the light emitted in this process depends on the type of atom or molecule which in turn depends on the energy of each electron and the depth to which it can penetrate the atmosphere. (We see that protons can play this game as well and produce their own unique red glow.)

We are now beginning to get a picture of the electron gun in nature's great auroral TV set in the sky, and maybe even a clue to the source of auroral substorms. Could they have something to do with the sporadic nature of the inflow of particles from the magnetotail? But what of the small-scale features of the aurora, features like individual rays, arcs and the folds in curtains? These details still elude space physicists.

Before returning to our tour, we might pause to ask 'what is the source of this mysterious electric field that, along with the Earth's magnetic field, seems to control the fate of the electrons and protons we have been watching?' This electric field spans the entire magnetosphere from dawn to dusk and from the magnetotail to the dayside magnetopause. It must be the product of a giant electric generator in space.

One of my most vivid memories as a boy was a toy electric generator. It consisted of a magnet in the shape of a 'C'. In the gap of the 'C', mounted on an axle, was a small loop of wire with many turns. The loop of wire could be rotated in the magnetic field by turning a small crank. The wire loop was connected to a miniature light bulb through slip rings that could carry electric current from the loop to the bulb. When the crank was turned the bulb would light. The faster the crank was turned, the brighter the bulb. Even more amazing, when the bulb was replaced by a small battery, and the loop given a little push, the loop would rotate by itself. The device became a miniature electric motor! I puzzled for hours over this.

How could turning the crank cause the bulb to glow? As I learned years later in a high school physics class, motion of the wires in the loop through the magnetic field would induce an electric field in the wires. This electric field, in turn, would drive a current through the bulb that would heat the filament until it was hot enough to glow. The key to the mystery is motion of the wire through the magnetic field of the magnet. The faster the motion, the higher the electric field, the higher the current, the hotter the filament, and the brighter the light. I did not know it at the time but, instead of turning the crank, I could have held the loop steady and rotated the magnet around and around and lit the bulb. The wire had to move through the magnetic field. When this happens an electric field is produced in the wire that drives an electric current. This is exactly what happens, on a much larger scale, in an electric power plant. Steam from coal, gas, or oil fires, or a nuclear reactor, or even sunlight, is used to 'turn the crank', giving us electric power in our homes and offices.

What has all this to do with the electric field we have discovered in the magnetosphere? Enter again the familiar solar wind!

Recall that the solar wind carries with it the magnetic field drawn out from the Sun, and the solar wind moves past the Earth at hundreds of kilometers per second. This means that the interplanetary magnetic field is being dragged over the magnetosphere at this speed. Because of this motion of the interplanetary magnetic field relative to Earth, a giant electric field is impressed on the magnetosphere. There are no wires in the magnetosphere but instead the near vacuum of space lets charged particles, like electrons and protons, move freely. So, just as in my toy generator, this electric field can drive currents and also move charged particles across the magnetotail toward the Earth during a geomagnetic storm. The strength of the electric field and its many resulting magnetospheric effects depends on the speed of the solar wind, and the intensity and direction of the interplanetary magnetic field. The electric field is strongest when the interplanetary magnetic field points southward. The reason for this will be obvious after the next stop on our tour.

We have seen that the solar wind acts on the Earth's magnetosphere like a giant electric generator that produces electric fields within the magnetosphere. These electric fields in turn drive currents that alter the configuration of the Earth's magnetosphere by the production of the ring current and by currents produced by the motion of electrons and ions along magnetic field lines that plunge into the atmosphere in the auroral zone. These 'field-aligned' currents are called Birkeland currents in honor of the famous Swedish scientist Kristian Birkeland whom we met earlier in connection with early investigations of the aurora.

The solar wind itself is also the source of at least some of the particles that this electric field moves into the magnetosphere to produce the ring current, Birkeland currents, the colorful aurora, and the Van Allen belts. How can we explain the entry into the magnetosphere of solar wind particles? The answer to this question is addressed at the next stop on our walking tour of the magnetosphere.

Let's walk quickly now from the night side to the day side of the magnetosphere. Once again, let's move outward away from the Earth toward the Sun. We pass the heart of the Van Allen belts, leaving behind the bulk of the ring current particles, with their dizzy gyrations and endless wild bounces between hemispheres. We approach the dayside magnetopause, the edge of the magnetosphere (tour stop 3, Figure 5.1).

At some point as we move outward, the magnetic field weakens and then suddenly vanishes altogether. We become dimly aware of a blizzard of very-low-energy electrons and ions at our back blowing toward the region of zero magnetic field. Beyond, we can see the solar wind and the interplanetary magnetic field all rushing toward us and into the zero field region from the other side. We notice that this onrushing magnetic field is pointing toward the south, opposite in direction to the Earth's field that we have just been passing through. The zero field region is the point where the two oppositely directed fields meet. It is called the 'magnetopause' because it is the very edge of the magnetosphere. At this point we see electrons and ions arriving from both sides, becoming accelerated or heated, and then shooting out in north and south directions at great speed.

The magnetic fields convected in from either side are merging and annihilating each other. Magnetic energy is disappearing and charged particles are capturing this energy for their own use; gaining energy. This process, taking place at the dayside magnetopause, is called 'magnetic merging' and it results in the acceleration of charged particles along the magnetopause, mainly to the north and south. Magnetic merging is believed to be a rather universal property in nature whenever magnetic fields with opposite orientations are brought together. It may be the major source of energy for solar flares and the acceleration of plasma near the equator in the magnetotail.

What is the fate of those particles caught up in the merging process and accelerated along field lines leaving the merging region toward the north and south? They become entrained on magnetic field lines that are connected to the Earth in the polar regions at one end and the solar wind at the other end. As they are pulled downstream by the solar wind, these magnetic field lines become draped over the poles much like a girl's long hair blown back over her head in a strong wind. Since the dayside merging and the sweeping downstream proceeds faster than the rotation of the Earth, magnetic field lines build up behind the Earth to form the long magnetotail, like a pony tail flowing in the wind. So the entrained particles, co-mingled solar wind and magnetospheric particles, from the dayside merging region, find themselves helplessly transported to the magnetotail.

When the solar wind and interplanetary magnetic field conditions are right for magnetic merging to proceed at the dayside magnetopause for an extended period of time, magnetic field and the accompanying charged particles thrown over the polar caps and into the tail may accumulate with no place to go. Space physicists believe that this accumulation results in the storage of magnetic energy in the northern and southern lobes of the magnetotail.

But the problem is how to release this accumulated energy. Magnetic merging springs to the rescue once again. Just as at the dayside magnetopause where two oppositely directed magnetic fields move toward each other, magnetic merging can also take place in the low or zero field zone near the magnetic equator. Here oppositely directed field lines from the north and south polar regions in the tail are pressed against one another by the continual flow of magnetic field lines over the poles (tour stop 4, Figure 5.1). This merging draws the field lines and their companion particles toward the equator and into the region where the great electric field can push them earthward once again. The only hitch is that the magnetotail magnetic merging process is not steady and continuous. Some process, not yet understood, triggers this merging so that it occurs periodically, each episode corresponding to the fresh injection of particles associated with a substorm. At last we have come a full circle and we see an explanation for the substorms that characterize the 'lights in the night'.

Our walking tour has given us a more detailed look at the endless cyclic processes that make the magnetosphere a fascinating circus of electric and magnetic fields, accelerating ions and electrons, and currents all in continuous but sometimes episodic motion.

CHAPTER EPILOGUE

The electric/magnetic/charged particle picture presented here may strike the reader as being at odds with the pressure gradient description of magnetospheric convection presented in Chapter 1. In fact, the two different physical descriptions are complementary and both are correct in their own right. One description treats the large-scale magnetospheric convection process as a fluid (through magnetohydrodynamics) and the other description looks at microscopic processes such as the motion of individual electrons and ions. Different physical forces are in play at the different scales.

6   The Sun: Where it all Begins

> The Sun is a mass of incandescent gas — a gigantic nuclear furnace where hydrogen is built into helium at a temperature of millions of degrees.
>
> Excerpt from the song 'Why Does the Sun Shine': lyrics by Hy Zaret, music by Louis Singer

It's all true. The Sun *is* a nuclear furnace where hydrogen is continually being converted to helium. The Sun devours about 6×10^{11} kilograms of hydrogen each second. Not all of the hydrogen reappears as helium, however. Some mass actually disappears in the process. Thanks to Einstein's famous $E=mC^2$ this mass is converted immediately to energy. About 5×10^{26} watts of heat energy are generated in this process. After a rather torturous path, this energy makes its way to the Sun's photosphere where it produces the incandescent ball of gas we see as the Sun. Ultimately, about 1,400 watts of sunlight fall on each square meter at the orbit of Earth.

What is less well known is that this nuclear burning takes place only at the innermost 25% of the Sun, the so-called core. So how does this heat generated in the core get to the outer layers of the Sun where it can shine on us? There are three ways heat can be transferred from one location to another: radiation, conduction and convection. Solar radiation is very familiar to us. It is the warmth we feel on our faces from sunlight. It comes to us in the form of visible and infrared electromagnetic radiation from the Sun. Heat conduction is also familiar. Put your hand on a hot stove and you have felt heat conduction from the direct contact between your hand and the stove. Heat transfer by convection is less familiar. Heat convection is the transfer of heat by the motion of material carrying heat from one location to another. A good example is boiling water. Bubbles that rise from the bottom of the pan carry hot air to the top. A second example, familiar to hawks and glider pilots, is the thermal updrafts of rising hot air that 'chimney' upward above an open field on a sunny day. Convection cells rise carrying heat upward until they cool then sink back down only to be reheated and rise again.

The intense heat generated by nuclear burning going on deep in the Sun escapes outward at first by scattered radiation, in the form of gamma-rays. This cooling by radiation can continue up to about 70% of the way to the surface. At this point, because of the cooler temperatures, the opacity of the solar gases is so great that radiation is no longer effective and further heat transfer to the surface must rely on convection. The outer 30% of the Sun consists of a layer of giant rising and falling convection cells. This continual churning motion can be seen (with appropriate filters, of course) in the visible portion of the Sun, the photosphere, as granulation (Figure 6.1). Granulation is a continuously changing pattern of light irregular spots surrounded by dark lanes. The light areas are rising hot gas where heat is brought to the photosphere and the dark areas are regions where the cooler gases are descending into the Sun to be reheated. This process heats the photosphere to about 6,000 kelvin and produces the incandescent yellow glow in the visible portion of the electromagnetic spectrum that we call the Sun. Here cooling by radiation takes over once again. We earthlings have developed sense organs that can make use of this outward flood of electromagnetic radiation. It is not an accident that our eyes are sensitive to the same portion of the electromagnetic spectrum where the solar radiation is most intense.

It is interesting that the electromagnetic radiation from the Sun's core, the temperature of which is about 16 million degrees, starts out as high-energy gamma-rays and ends up as visible photons from the 6,000 degree photosphere. Quite a remarkable cooling feat for the Sun!

But what has all this to do with storms in space? The photosphere and its convection are responsible for our space weather. Through processes we don't fully understand, the photosphere heats a layer of gas above it, the corona, to several million degrees. As we have already seen, the high temperature of the corona is the source of the solar wind.

Because of the low density of the corona and the much brighter photosphere beneath it, the corona can only be seen in the visible

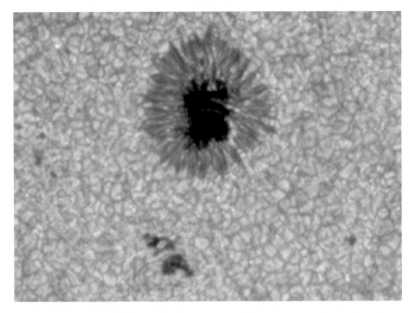

FIGURE 6.1 A close-up look at a sunspot. Sunspots are regions where the solar magnetic field becomes very intense and bubbles up from below the photosphere. The sunspot is the dark region (umbra) surrounded by the radial penumbra. Sunspots may be 50,000 kilometers across and can last for months. The mottled region covering the remainder of the photosphere is called granulation and is due to the 'boiling' of hot gases rising up from below to carry heat from the interior of the Sun. Source: Data courtesy of P. Brandt (Kiepenheuer Institut für Sonnenphysik, Freiburg, Germany), G. Scharmer (Uppsala, Sweden) and G. Simon (National Solar Observatory). Primary image processing by D. Shine, Lockheed Corporation; The High Altitude Observatory Archives.

portion of the electromagnetic spectrum at the edge of the Sun when the photosphere is blocked from view during an eclipse or using a coronagraph. When viewed edge-on the corona is seen to extend high above the photosphere (Figure 2.1). Because of its extremely high temperature, the corona can, however, be seen on the disk of the Sun in the X-ray portion of the spectrum. Ordinarily the X-ray Sun is hidden from earthlings by the absorption of X-rays by the Earth's atmosphere. During the famous SKYLAB missions following Apollo, astronauts obtained the first extensive photographs of the X-ray Sun.

FIGURE 0.1 Artist's conception of the International Space Station approaching dawn. Source: NASA.

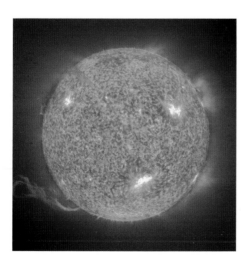

FIGURE 0.2 Image of the Sun in extreme ultraviolet wavelengths taken from the extreme ultraviolet imaging telescope aboard the SOHO spacecraft. Source: Solar & Heliospheric Observatory (SOHO). Data courtesy of SOHO/EIT consortium. SOHO is a project of international cooperation between ESA and NASA.

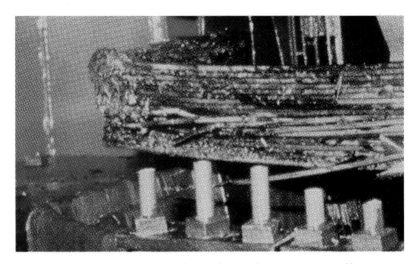

FIGURE 0.3 A power transformer damaged by currents created by a magnetospheric storm. Source: NOAA/Space Environment Center.

FIGURE 1.1 The solar corona as seen during the solar eclipse photographed in India on February 16, 1980. Source: Image digitized from photographs in the archives of the High Altitude Observatory. Primary image processing by A. Stanger, High Altitude Observatory, National Center for Atmospheric Research (Boulder, CO).

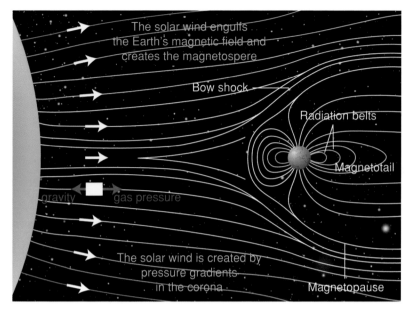

FIGURE 1.2 Artist's conception of the solar wind created in the solar corona and surrounding the Earth's magnetic field to form the magnetosphere. The magnetosphere and distance to the Sun are not to scale. Modification by the author of a drawing from Y. Kamide (used by permission).

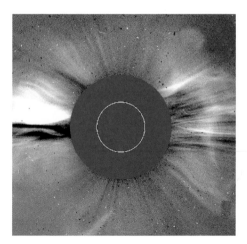

FIGURE 2.2 In this artificial coronagraph picture, the LASCO solar image camera aboard the SOHO spacecraft has captured a coronal mass ejection headed straight for the Earth. The CME has expanded so that it is larger than the Sun. The Sun is blocked out by the dark disk in the center to render the CME visible. Source: Solar and Heliospheric Observatory (SOHO). Data Courtesy of SOHO/LASCO consortium. SOHO is a project of international cooperation between ESA and NASA.

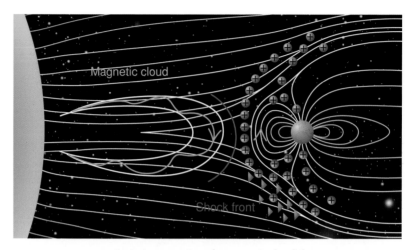

FIGURE 2.3 Artist's conception of a magnetic cloud about to encounter the Earth's magnetosphere. The magnetic cloud is preceded by a bow shock wave (red) and contains entrained within it the twisted remnant of the solar magnetic field dragged out from the Sun (orange). Note that the orientation of the magnetic cloud field is opposite to that of the Earth's magnetic field, a favorable condition for magnetic merging at the magnetopause. The red dots represent energetic protons that have been accelerated to high energies at the bow shock wave and are able to penetrate the magnetosphere in the polar regions. A few energetic electrons, triangles, can be seen as well. Modification by the author of a drawing from Y. Kamide (used by permission).

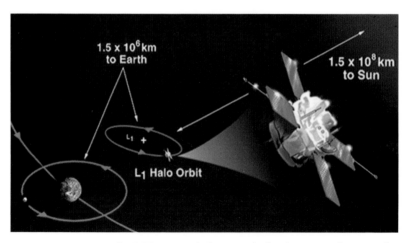

FIGURE 3.1 The ACE spacecraft shown parked at the L1 point between the Sun and the Earth. Source: NASA.

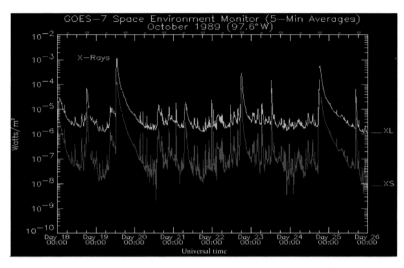

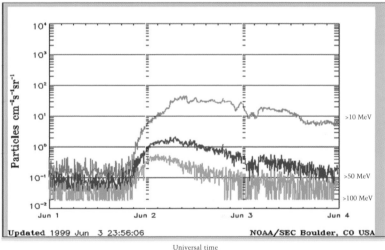

FIGURE 3.2 The top panel shows a sample of solar X-ray data taken by the GOES-7 spacecraft. The sharp upward spikes indicate X-rays from solar flares. X-rays in two energy ranges are shown. The bottom panel shows the GOES –7 solar energetic proton flux in three energy ranges from a single solar storm. Notice that the flux remains high for more than two days. Source: NOAA/Space Environment Center.

FIGURE 4.1 The aurora over Isle Royale National Park, Lake Superior.
Source: Dan Urbanski, Silver Image Studios, Silver City, Michigan, USA.

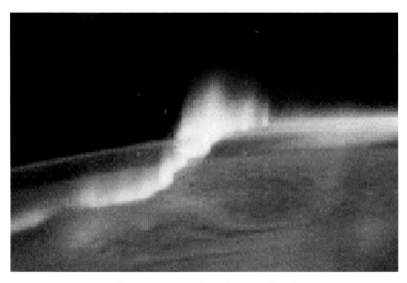

FIGURE 4.2 The aurora as seen from the space shuttle Discovery.
Source: NASA.

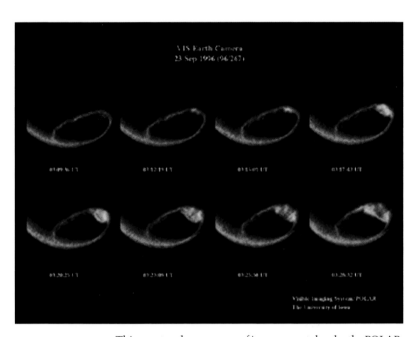

FIGURE 4.3 This spectacular sequence of images was taken by the POLAR satellite from high above the north pole. It shows the entire auroral oval during a substorm. The sunlit side of the Earth is the arc to the lower left. The auroral oval is the circle surrounding the pole and the bulge that grows on the circle to the upper right is the auroral substorm. The images are taken about three minutes apart. This ultraviolet image was acquired with the Earth camera that is one of three cameras in the Visible Imaging System (VIS). The design and assembly of the VIS was performed by the VIS team at the University of Iowa. The VIS is one of twelve instruments on the POLAR satellite of the NASA Goddard Space Flight Center. The Principal Investigator is Dr. L.A. Frank and the Instrument Scientist and Manager is Dr. John B. Sigwarth. This image was provided courtesy of Dr. Louis A. Frank at the University of Iowa.

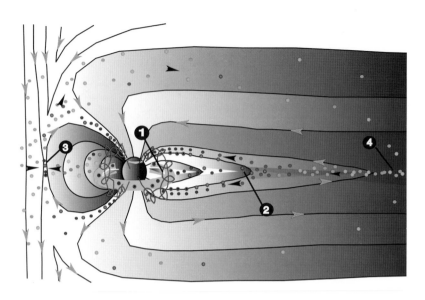

FIGURE 5.1 The roadmap for our walking tour of the magnetosphere. The tour stops are: (1) the inner magnetosphere, home to the trapped particles in the Van Allen radiation belts and the ring current particles; (2) the near-Earth magnetotail or the inner edge of the plasmasheet where particles enter the inner magnetosphere and then feed the ring current or move along magnetic field lines to precipitate into the auroral zone; (3) the dayside magnetopause where magnetic energy is converted to particle kinetic energy and magnetospheric convection begins when the interplanetary magnetic field has an orientation opposite to that of the Earth's magnetic field; and (4) the neutral sheet of the magnetotail where further magnetic merging may take place to accelerate charged particles Earthward and create magnetospheric storms. This figure also shows the circulation of particles from the solar wind into the magnetosphere at high latitudes over the poles, into the magnetotail, and finally Earthward near the center of the magnetotail. Protons are yellow at low energies then pink and red at higher energies. Electrons are blue –light blue at low energies and darker blue at higher energies. The dark arrows show the direction of motion of the particles. The ring current can be seen as a light blue ring encircling the Earth. Protons move clockwise around the Earth and electrons move counterclockwise. The red and blue helical particles represent energetic protons and electrons in the Van Allen belt. (Not shown is a second Van Allen belt of even more energetic electrons and protons closer to the Earth.) The auroral zone is shown as a yellow circle around the pole. Black lines represent magnetic field lines and the pink arrows are the direction of these lines.

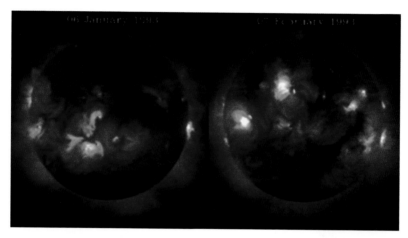

FIGURE 6.2 The Sun as seen in X-ray wavelengths on two successive days. Bright regions are 'active' regions with temperatures in excess of 2 million degrees. The dark regions are cooler regions called coronal holes. The solar wind that emerges from coronal holes has a higher average speed than the rest of the solar wind. Source: The Japanese satellite Yohkoh (the Yohkoh Science Team), the High Altitude Observatory Archives, High Altitude Observatory, National Center for Atmospheric Research (Boulder, CO).

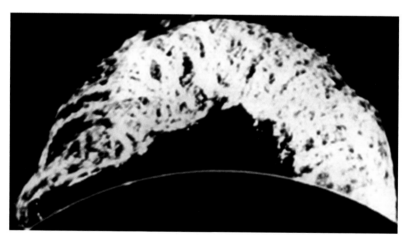

FIGURE 6.3 A giant prominence extends out from the limb of the Sun. This eruptive prominence was photographed at the H alpha wavelength in June of 1946. It has a height of 200,000 kilometers above the photosphere. One of the largest ever observed from the ground, it is called 'Grand Daddy'. Source: Image digitized from a photographic print in the archives of the High Altitude Observatory, National Center for Atmospheric Research (Boulder, CO).

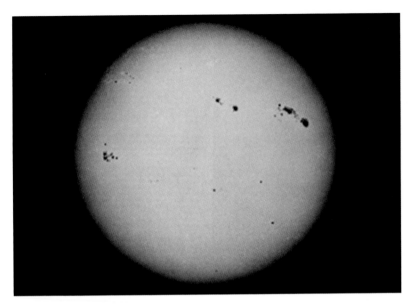

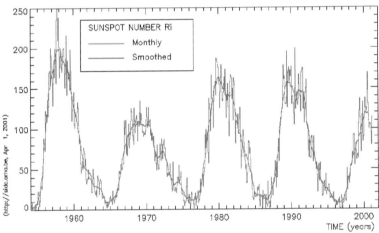

FIGURE 6.4 The top panel shows several sunspot groups on the disk of the Sun. The bottom panel shows the monthly average sunspots over the last four solar cycles. The 11 year sunspot cycle can easily be seen. The current sunspot maximum occurred around the middle of 2000. Source: Eclipse photograph digitized and processed by A. Lecinski, High Altitude Observatory, National Center for Atmospheric Research (Boulder, CO) (top panel). Data prepared by the Royal Observatory of Belgium (bottom panel).

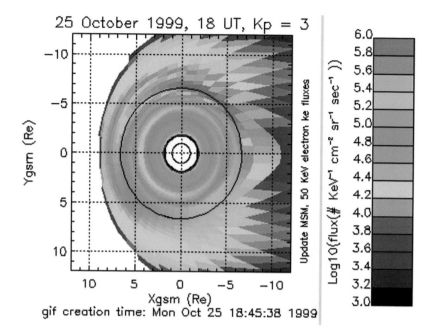

FIGURE 7.1 A space weather map: the Magnetospheric Specification Model as shown on the NOAA Space Environment Center web page. The color-coded contours represent the intensity of electrons near the equatorial plane of the magnetosphere. The Earth is the small circle in the center. The larger outer circle represents the geostationary orbit where several hundred communication, weather, navigation, and surveillance satellites are orbiting. A magnetospheric storm is in progress and the bright orange and red region that lies along the geostationary orbit, near the top of the circle, indicates a hazardous zone for satellites where the electron intensity is high. The Magnetospheric Specification Model is driven by continuous data on solar wind conditions from the ACE spacecraft and by data from ground-based magnetic observatories processed by the US Air Force in real-time. This map is updated by NOAA every hour. Source: NOAA Space Environment Center. The Magnetospheric Specification Model was developed at Rice University for the US Air Force Research Laboratory.

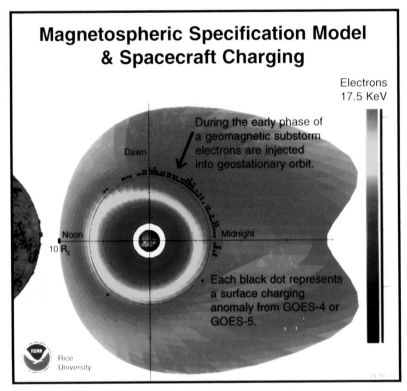

Magnetospheric Specification Model & Spacecraft Charging

Electrons
17.5 KeV

Dawn

During the early phase of a geomagnetic substorm electrons are injected into geostationary orbit.

Noon

10 R_E

Midnight

Each black dot represents a surface charging anomaly from GOES-4 or GOES-5.

Rice University

FIGURE 8.1 An overlay of the output of the Magnetospheric Specification Model and the locations of two of the geostationary orbit GOES spacecraft at the time they exhibited operational anomalies resulting from charging of the surface of the satellites by electrons. It can be seen from this figure that most of the GOES anomalies occur in the region of space where the storm-generated electron flux is the highest. This is evidence of the link between the space environment and satellite operational problems. Source: Dan Wilkinson, NOAA National Geophysical Data Center and the Department of Physics and Astronomy, Rice University.

These X-ray images reveal a totally new face of the Sun. The corona is seen to be highly variable with regions of intense disruptions associated with active regions in the photosphere (Figure 6.2, color section). There are even places where the corona is almost missing altogether, called coronal holes. Surprisingly, coronal holes are the source regions for a higher-speed solar wind. The variability of the corona gives rise to variability in the solar wind. Active regions can produce great prominences and CMEs that drive our magnetospheric storms (Figure 6.3, color section). But, even during times of low solar activity, important magnetospheric storms arise from the interface that develops between the quiet solar wind and the high-speed solar wind that emerges from coronal holes.

We cannot overlook another important feature of the interior of the Sun: the magnetic field. Five billion years ago, as the Sun was forming from a giant molecular cloud in interstellar space, the weak magnetic field that threads through all of space became caught up in the material of the gathering solar nebula. As the nebula drew together to form the Sun at the center, it began to rotate and at the same time the entrapped magnetic field intensified. This rotation and contraction of the solar nebula resulted in the creation of electric currents within the young Sun that produced further amplification of the magnetic field through a dynamo-like process. This field, which must have pervaded the entire solar nebula, probably provided the 'seed' magnetic field from which the Earth's dipole magnetic field arose.

Indeed the Sun's magnetic field would probably be a giant dipole field were it not for the solar wind and irregularities generated by transient active regions on the Sun. The solar wind is highly conducting and carries a great deal of momentum. It is able to drag the Sun's field outward and distort it from a true dipole, particularly near the equator. Near the poles of the Sun the field is more nearly dipolar. We know that the direction of this dipole reverses every 11 years. The reversal of this field is synchronous with levels of sunspot activity.

The Sun rotates faster near the equator than at the poles. This 'differential rotation' causes magnetic field lines within the Sun to be wound into two tight spirals, much like string on a top. The field spiral winds in opposite directions on either side of the equator. As the spirals grow tighter, the magnetic field increases and at some point the strength of the underlying magnetic field overpowers the overlying solar photospheric gas and bursts through the photosphere to form loops or arches of magnetic field that stick out above the photosphere. Where the magnetic field protrudes through the photosphere the photospheric gas is pushed aside forming cooler, dark regions known commonly as sunspots (Figure 6.1 and Figure 6.4, color section). Because two sunspots form from a single loop of magnetic field lines, they tend to travel in pairs with the field emerging from one sunspot and returning to the Sun at the other. The two sunspots are said to be of opposite magnetic polarity. The leading sunspots of each pair in a given hemisphere of the Sun will tend to have the same polarity and that polarity will be opposite in the two hemispheres. It is even more remarkable that the polarity of the leading sunspot reverses every 11 years. This means that the direction of the Sun's dipole magnetic field is undergoing a full reversal every 11 years! (The same effect is known to take place for the Earth's magnetic field but over a period of several hundred thousand years.) Moreover, the number of sunspots on the Sun varies over a period of 11 years, with new sunspots appearing at mid-latitudes on the Sun and slowly migrating toward the equator. The motion of sunspots across the Sun and reversal of the direction of magnetic field within sunspot pairs define the well-known 11-year solar cycle (Figure 6.4).

The area around sunspot pairs or groups frequently becomes the site of intense activity, often resulting in great brightening called a solar flare. As we have seen, solar flares can be the source of intense radio noise, X-ray fluxes and solar energetic protons that can produce problems with radio communication through changes in the Earth's ionosphere. The probability of flares follows the sunspot cycle but flares can occur at any time during the cycle.

Until recently it was believed that solar flares were chiefly responsible for changes in the solar wind that produce magnetospheric storms. This dubious honor is now being bestowed on coronal mass ejections. CMEs frequently arise from the area where a flare may appear; however, they are not always associated with flares.

Coronal mass ejections appear as a bulging or arcade in the corona high above the intense magnetic fields of a photospheric active region (Figure 2.1). It appears that the energy that drives flares and CMEs derives from the intense, twisted magnetic fields emerging from the photosphere at sunspots. How this energy is released is not yet fully understood but probably is related to the merging of oppositely directed field lines as we have proposed takes place at the Earth's magnetopause and in the magnetotail. Magnetic energy is transformed into particle flow energy and the CME is blasted away from the Sun.

The looping coronal prominences that burst outward to become CMEs carry twisted solar magnetic fields to the Earth's magnetosphere. The periodic southward orientation of these twisted fields provides for maximum coupling with the Earth's magnetosphere through magnetic merging at the edge of the magnetosphere on the day side during passage of the magnetic cloud. Of course, the effect of a CME will not be felt at Earth unless it is launched from a point on the Sun where its trajectory through space will intersect the Earth's magnetosphere. The effect will be maximized if the Earth is in the center of the bull's-eye.

We now have a clearer picture of the complete chain of events that starts with the nuclear fusion at the core of the Sun and ends with a giant storm in the Earth's magnetosphere.

The Earth is not the only planet subject to the Sun's mischief. Jupiter, Saturn, Uranus and Neptune, and even tiny Mercury, are known to have magnetospheres. Jupiter has very intense radiation belts, magnetospheric storms and auroras. Venus and Mars have magnetic fields too weak to sustain magnetospheres. We know very little about tiny Pluto but, even at the great distance of nearly 40 times the orbit of Earth, it is within the reach of the solar wind. Traveling at high velocities, the solar wind has sufficient momentum to carry it to Pluto and beyond. All the planets are within the sphere of influence of the Sun's extended atmosphere, the solar wind. In fact, our most distant spacecraft, Voyagers 1 and 2, now well beyond the orbit of Pluto, have not yet reached the outer limits of the solar wind. As it moves toward the outer reaches of the solar system, the solar wind gradually slows as it climbs the steep gravitational hill produced by the Sun. At some point the solar wind pressure will fall to match the pressure of the gas in the interstellar medium through which the Sun is moving. At that point, the heliopause, the solar wind will lose its outward motion and its influence will end at last.

7 Nowcasting and Forecasting Storms in Space

It's hard to make predictions—especially about the future.

It is nearly midnight. I am putting the finishing touches on the lecture for my 9 AM class when the phone rings. My brother-in-law is calling from Milwaukee with the news that there is a brilliant auroral display going on. It is rare for the aurora to be seen that far south of the auroral zone. He knows of my work on space weather and that our research group has supplied the Air Force and NOAA with computer *models* that indicate the status of storms in space. He wonders if our models are showing indications of any unusual magnetospheric activity.

Delighted at the chance to catch a storm in progress, I put down the phone and quickly start my computer, launch the Internet browser, and open the NOAA, Space Environment Center (SEC) *Magnetospheric Specification Model* (MSM) real-time web page. A color contour plot much like a conventional temperature weather map pops onto the screen. Instead of temperatures across the US, this space weather map shows the intensity of energetic electron fluxes in space on a slice through the equatorial plane of the magnetosphere (Figure 7.1, color section). Immediately I realize that the Magnetospheric Specification Model is showing a very intense storm in progress. There is an orange and red area that stretches in an arc from midnight around almost to noon that symbolizes a danger zone of intense electron flux. I also notice that the indicated pseudo Kp index is 7.6 showing that a strong storm is in progress. This is the largest storm I have seen since our model came on-line at SEC several months ago.

Picking up the phone, I report that the Magnetospheric Specification Model indicates an intense space storm in progress. Later, it's back to the computer to check the next 15-minute update from the MSM. A few email messages are sent to colleagues around the world to alert them to the storm that still appears to be picking up steam. I check the routine space weather alerts from the NOAA/SEC forecast center but there is no warning posted yet. The NOAA forecasters there have not yet noticed the gathering storm.

The next morning I tell my solar system class about the storm. We discuss storms in space and how the Magnetospheric Specification Model is able to 'nowcast' the progress of a storm.

After class it's back to the Internet to follow the progress of the storm. By now the word is out and Andrew Urquhart, a graduate student in the Rice University Department of Space Physics and Astronomy, is busy collecting data about the storm from a variety of sources, particularly the space weather stations. Inspection of ACE data indicates that the solar wind speed reached over 900 kilometers per second and the IMF attained higher than normal values and was southward for a long time. We can identify the X-rays from a flare that occurred a day earlier—probably related to our big geomagnetic storm. We also find evidence of a strong burst of solar energetic protons in the geostationary satellite data.

This true story illustrates some aspects of real-time space weather nowcasting through the use of computer models. Because of the problems storms in space can cause for man-made systems, both in space and on the ground, it is important to know what is going on in space nearly everywhere, all the time. Space weather monitoring satellites are useful for telling what the space weather conditions are at certain locations in space, but we need the bigger picture. Magnetic indices such as Kp and Dst give us a measure of the overall storm intensity. They cannot, however, tell us conditions at specific points in space such as the flux of energetic electrons at the location of a single communication or navigation satellite or the ionospheric currents and magnetic disturbance near an electric power station. What is needed is a means of extrapolating from the satellite data samples taken at a very few isolated points in space to a determination of storm conditions at all locations in the magnetosphere. This extrapolation can be accomplished by computer models that incorporate the basic physical processes believed to drive the storm.

A 'model' is a mathematical simulation that provides a large-scale representation of the conditions at all points within the model, based on the physical conditions that are believed to cause the storm. Usually a weather model will be driven by a continuous stream of input data from monitoring stations that report conditions at certain locations. Tropospheric weather models, for example, take temperature, wind vectors, pressure and humidity data from various weather stations and, using the laws of physics, compute what the temperature etc. will be at other locations. Usually these locations are on a grid that covers the region within the boundary of the model. Models of this type are possible only because high-speed computers can keep track of the data and perform the computations fast enough to be useful. Space weather models perform a similar function for conditions in space using input data from the space weather stations that we have already learned about. An example of the output of interest for space weather models is the intensity of energetic particles in the magnetosphere.

The Magnetospheric Specification Model is one such space weather model. The MSM was developed at Rice University starting in the late 1980s, with the support of the US Air Force. It took nearly ten years for a team of scientists and graduate students to design, build, and test the MSM. The MSM computes the intensity of energetic electrons and ions in the equatorial plane of the Earth. It uses models of the Earth's magnetic and electric fields developed specifically for MSM to trace the motion of the electrons and ions. The MSM uses data that come from satellites and ground magnetometers to continually adjust the model parameters. These data can be fed to the MSM as it is running in real-time. This provides continuous model output that gives the conditions of a storm as it is actually going on. This is called 'nowcasting'. Nowcasting is useful because it provides a picture of where the most intense features such as high-energetic particle fluxes are at any time, and of how the storm is changing; just as tracking the location of a hurricane gives us an idea of where it is and where it is headed.

The MSM can also be run with input data from a past storm to determine how severe the storm was at the location of a satellite. This 'retrospective' analysis can be useful to help diagnose a problem with a satellite that may have experienced an operating anomaly.

While models that can provide information on space storms as they are occurring are useful, just as with terrestrial weather, we would like very much to be able to predict when a storm is coming. A goal of space weather research is to develop forecast models. As with terrestrial weather, the longer the forecast the better.

Forecast models are more difficult to build than retrospective or real-time models. Any model requires a continuous stream of information about the changing conditions at the boundary of the region of space for which the model is being run. For example, in the MSM we have an outer boundary of the model. We must have a way of knowing, or specifying how conditions are changing on that boundary. The energy source moves inward across that boundary to feed the storm. In the case of the MSM, the energy source at the boundary is obtained from statistical models that are themselves driven by real-time data like the Kp index. To convert the MSM to a forecast model, we need a way to determine the boundary conditions in advance, ahead of real-time. We cannot wait for data to come in from ground magnetometer weather stations from which we can compute Kp.

To use an analogy from terrestrial weather, imagine a hurricane approaching the Gulf of Mexico from the Atlantic Ocean. The job of a forecast model is to determine where it will go once it enters the Gulf. We might have a model of the Gulf with a roughly circular boundary located at the land–sea boundary for the Gulf. The track of the hurricane will be determined by barometric pressures throughout the Gulf and by winds entering the Gulf from all sides. In order to know the future track of the hurricane, we would like to know the pressures and steering currents not just at the present time but also in the near future. This requires a forecast of the boundary conditions of the model. These may be obtained by still other models which deal with the surrounding areas, particularly those in the direction from which the weather is coming.

In the case of the MSM we have taken a unique approach. *Artificial neural networks* have been trained to forecast the boundary conditions. These forecasted boundary conditions can then be used to drive the model in the same fashion as for a nowcast but the model output now becomes a forecast.

Artificial neural networks (ANNs) are computer imitations of the way biological neurons allow humans to learn and predict future conditions. Suppose a friend introduces you to someone at a party. You talk with the person for a while and soon become familiar with their appearance and the sound of their voice. Some days later, perhaps at another party, from across the room you hear the familiar sound of the same voice. Without actually seeing the person, you know who they are and what they look like. At the pervious meeting, the neurons within your brain have trained themselves to associate the sound of that voice with the person's appearance and other characteristics. Your neurons can even predict the future! If you stroll across the room to once again meet your new friend, you will know what he or she looks like before you actually cross the room. In a similar fashion, inputs to mathematical neurons within a computer can be trained to recognize and associate combinations of inputs with certain outputs. This training is accomplished by the repetitive presentation of input data and simultaneous observations while the artificial neurons adjust themselves by changing the strength of connections between neurons until the relationship is established. The neurons learn to identify patterns in the input data during training and when presented with similar patterns during a set of new input data they replicate the type of output they have learned to identify with that input during training. The success of this method depends on the quality of the data used for training and the skill of the scientist doing the training. Obviously, there must be some relationship between the input data and the output. An important aspect of using neural networks involves discovering which input parameters are important. Neural networks are being used today for a variety of tasks such as controlling chemical processes, determining if or when a business is going to fail, and even whether or not an applicant is a good credit risk.

The forecast version of the Magnetospheric Specification Model that uses artificial neural networks to forecast the input parameters is called the Magnetospheric Specification and Forecast Model (MSFM). This model can provide good one-hour forecasts of magnetospheric conditions; longer forecasts are available with lower accuracy. An advanced version of the MSFM uses data from the ACE spacecraft. As we have seen, ACE is nearly an hour upstream in the solar wind. This upstream lead time makes possible an even longer forecast.

As with terrestrial weather models, for space weather the ultimate issue is accuracy. It is a fair question to ask: 'how good are the models?' A related question is: 'how good are the measurements upon which the models are judged?' Measurements of space weather parameters are difficult to make with high accuracy because the instruments must be left alone in space for long periods of time. These instruments are subject to drift and degradation by the very environment they are sent to measure and generally cannot be retrieved after launch for rechecking. Furthermore, it is difficult to simulate the exact space plasma environment in the laboratory and so calibration is difficult. Some satellite instruments are even launched without preflight calibration, usually because of lack of funds or the failure to appreciate the importance of calibration. Measurements of the flux of electrons in a particular orbit in space have been known to vary substantially when made by similar instruments on two different satellites.

The question of model accuracy is also different from the standpoint of the model developer, the space weather forecaster, or the satellite or power station operator. The model developer is primarily interested in whether the parameters specified by the model agree with those that would be measured in space. For example, based on testing against a variety of magnetospheric storms, the MSM predicts electron fluxes which lie within three times the values measured by satellites. The model is somewhat better for protons. This is probably comparable to the accuracy of the satellite measurements themselves. The MSM does better at some locations in space than others and there are some features of storms for which it does not provide a good simulation.

On the other hand, the professional governmental space weather forecaster may be interested in larger issues such as the hit or miss accuracy of forecasting storms; if the model forecasts a storm, did one actually occur. Because of his/her concern regarding storms, the space weather forecaster is particularly interested in the forecast of storm indicators like Kp and Dst. For this reason NOAA's Space Environment Center in Boulder, Colorado, encourages the development of models for the prediction of storm indices. NOAA will judge models based on the forecast reliability, or skill score.

Commercial satellite operators and power companies have a more narrow focus regarding space weather forecasting. They are primarily interested in having some warning that conditions are ripe for the disruption of service precisely at the location of the operator's satellite or power plant. This is where computer models like MSM, that can give detailed information at every location, become important. Tailoring information from the models to the needs of individual customers may go beyond the capabilities of NOAA. This has opened the door for commercial space weather forecasters who provide value-added, customized services to government-run model output. Moreover, the conditions that may cause problems often differ for different systems. Like humans, different satellites have different sensitivity levels for pain. Each operator must know his own system and may need a team of engineers who understand the effects of a given space condition on his system as well as the accuracy of the forecasting model.

The MSM and MSFM are first-generation space weather models. They are quite good but they are based on submodels of the electric and magnetic fields that are built more on experience that on first principles. MSM and MSFM do not cover the full range of space and particle energies. More sophisticated models are currently under development at several universities and corporations. These newer models, called magnetohydrodynamic or MHD models, employ internally self-consistent electric and magnetic fields and automatically couple the ionosphere and magnetosphere to provide a much more complete picture of the storm than the MSM. Unfortunately, the new models also require more powerful computers and are much slower than the MSM. Even MHD models have some shortcomings. Since they treat the space medium as a fluid, they do not incorporate the correct physics for the motion of individual energetic particles. It is necessary to add certain particle submodels into the MHD models. For example, submodels can be overlaid onto the MHD models to treat energetic particles, the radiation belts, and the solar energetic particles. The magnetic and electric fields from the MHD model can be used in the individual particle models to obtain a more accurate specification or prediction. These combined models are called hybrid models. When they are fully developed and the computer power is available to run them in real-time, hybrid MHD models of the magnetosphere will no doubt become the second-generation space weather models.

Up to this point our discussion of forecast models has focused primarily on models of the magnetosphere and forecast times of the order of an hour or so. But, as we have seen, the Sun and the solar wind are the source and means by which space storms come to us. The US Air Force, among others, has a requirement for 24 hour or longer forecasts of space weather. Any long-term forecasts, greater than a few hours, must involve observations of the Sun and models of the propagation of the solar wind. We can envision complete, coupled models that will cover the entire space weather process from the Sun's photosphere all the way to the Earth's ionosphere. In fact, plans for such models are already underway at the University of Michigan, Boston University and other places. We can look forward to the day when TV weather reports will include the weekly space weather forecast.

One important aspect of sophisticated models, aside from their potential contribution to space weather forecasting, is that they challenge the frontier of high-speed, numerical computer modeling. Many applications for the type of modeling done for forecasting storms in space can be found in industry. There are many similarities between the numerical models I have described and those used, for example, in exploration geophysics or chemical plant process control. Graduates who have done space physics modeling have gone on to find jobs in the petroleum industry or similar industrial or governmental areas.

Worldwide there is currently a frantic race to design and build faster computers. Development of high-speed computing capability requires development of modeling software capability as well as hardware development. The race to build faster computers will not be meaningfully fulfilled unless sophisticated models that can take advantage of the speed are also funded and developed. The investment in space weather models pays dividends in other areas as well.

We have also seen how spacecraft observations play a key role in real-time forecasting of space storms. It is also true that without long-term routine space measurements the development of forecast models would be impossible. Space physicists who build models must have extensive and detailed measurements of the space environment in order to understand the complex physical processes that govern the space environment. Many of these processes are not yet understood after nearly 40 years of space exploration. This process has been slow and tedious for several reasons. The magnetosphere and solar wind cover a vast amount of space and individual spacecraft are able to make measurements at a single isolated point. This means that one or two satellites give a very incomplete picture of the 'whole animal' at any one time. It is a bit like the seven blind men examining an elephant—only there are fewer than seven men, the elephant is 200,000 kilometers across, far more complex, and continually changing. This problem has been reduced somewhat with the recent launch of a new spacecraft capable of remote sensing almost the entire magnetosphere, the IMAGE satellite. IMAGE stands for Imager for Magnetopause-to-Aurora Global Exploration. IMAGE makes use of new particle sensing technology. With IMAGE we are able for the first time to get the big picture—to see what is going on in the magnetosphere everywhere at the same time.

Model development is also made difficult by the 11-year solar cycle discussed in Chapter 6. We have only experienced four complete 11-year cycles since the start of the space age. We are forced to watch the Sun change in slow-motion.

Despite all of these problems, as we have seen, progress has been made in understanding the causes and mechanisms of storms in space and crude models have been built that can provide nowcasts and even short-term forecasts. An infrastructure exists to disseminate information obtained from real-time space weather stations and the models. Located in Boulder, Colorado, the NOAA Space Environment Center maintains a round-the-clock forecast staff that continually monitors the Sun and data from space weather satellites and ground-based observatories. SEC issues regular bulletins indicating current and forecasted space weather conditions. They provide alerts and warnings of impending hazardous space conditions. Their customers include a large range of commercial, government and even private users of space weather data. Commercial users and individuals can access the SEC web site (sec.noaa.gov) and obtain the latest forecast, alerts and warnings as well see the real-time data output from solar wind monitors and the MSM. The US is not alone in the space weather business. Providing data is a cooperative international effort. Japan maintains a similar space weather forecasting facility and other countries such as Sweden and Finland are developing space weather forecasting stations.

The US Air Force has been a leader in space weather observations, and research, and has supported the development of space weather forecasting models. The Air Force, in addition to maintaining a space weather forecasting facility of its own and cooperating with NOAA at SEC, also has a fleet of satellites that make space weather measurements. This fleet of satellites has provided valuable data for space weather model development. Without the long-term, reliable data provided by the Air Force and Department of Energy satellite systems operated by Hanscom Air Force Base and Los Alamos National Laboratory, models such as the MSM could not have been developed. NOAA also operates space weather geostationary orbit monitoring satellites. NASA operates the ACE and IMAGE satellites and has, over the years since the start of the space age, provided many research satellites that have helped to unlock many of the mysteries of the magnetosphere, the ionosphere, the solar wind and the Sun.

Investigation of the space environment has been an international endeavor from the beginning and many nations have contributed numerous spacecraft to the effort.

8   Technology and the Risks from Storms in Space

> 'Space Weather' refers to conditions on the Sun and in the solar wind, magnetosphere, ionosphere, and thermosphere that can influence the performance and reliability of space-borne and ground-based technological systems and can endanger human life or health.
>
> *The National Space Weather Program Strategic Plan*, August 1995

Around Alone must be one of the most arduous athletic contests ever devised. It is also a proving ground for the latest in space-age electronics, marine and materials technology. *Around Alone* is a single-handed sailing race around the world. Run in four legs that take the better part of a year to complete, *Around Alone* challenges solo sailors with extremely severe open ocean weather and incredible physical and psychological conditions.

In February of 1999 leg three of *Around Alone* was underway between Auckland, New Zealand and Punte Del Esta, Uruguay. This leg takes competitors through the Southern Ocean, a stretch of the far South Pacific that circumnavigates Antarctica. The weather conditions are formidable and the nearest land can be several thousand miles distant, far beyond the range of normal search and rescue operations. Westerly currents combined with very strong, west tailwinds make for fast times but provide particularly treacherous sailing conditions.

Nineteen hundred miles out of Auckland, Frenchwoman Isabelle Autissier's boat, *PRB*, capsized when caught by a sudden wind shift. The mast broke as the sail was dragged beneath the waves. What happened next is an example of modern space technology on several fronts.

All *Around Alone* competitors carry several types of emergency radio beacons. The newest of these is a GPIRB (Global Position Indicating Radio Beacon). Isabelle's GPIRB was activated immediately and began transmitting a distress signal identifying her boat and describing her exact location. The GPIRB uses the famous Global Positioning System (GPS). GPS employs a constellation of satellites to transmit radio signals that allow a user on Earth or in the air to determine his or her position with high accuracy.

Isabelle's GPIRB emergency signal was intercepted by the search and rescue satellite system, COSPAS-SARSAT/GEOSAR. This system uses geostationary weather satellites (GOES) and low-altitude satellites for monitoring search and rescue beacons. The distress call was relayed by satellite to SARSAT command posts.

Around Alone skippers keep in touch with the Race Operations Center, ROC, by satellite communication systems, COMSAT-C. COMSAT Corporation is a sponsor of *Around Alone*. The COMSAT-C communication system uses a system of geostationary satellites to provide email, digital photo, and weather communication for the race competitors. Skippers are required to provide regular GPS-based position reports every 6 hours via COMSAT-C. Email is the main means of communication between the racers, their shore crews and ROC. Earlier in the race a Russian skipper performed life-saving surgery on himself by following instructions from a doctor sent to him by email. Satellite phones are also used for communication. Solar cell arrays mounted on deck provide power for the onboard electronics.

When SARSAT operations received Isabelle's distress call they immediately notified ROC that her GPIRB had been activated. Italian competitor Giovanni Soldini's boat, *FILA*, was then 200 miles behind *PRB* to the north and west. Believing that he was in the best position, Race Operations Center advised Soldini by email of Isabelle's situation and he immediately set a course to attempt a rescue. It took Soldini 24 hours battling strong winds to reach the location indicated by Isabelle's GPIRB. After a brief search, Giovanni found Isabelle's overturned boat within a few miles of the point indicated by the GPIRB's initial report, but she was not to be seen. Giovanni, suspecting she was in the boat made one more pass and threw a hammer at the hull of the boat. Isabelle emerged through an escape hatch at the stern, and an amazing rescue was accomplished. Under race rules Soldini was able to rejoin the race with Isabelle onboard but not assisting with the sailing. A race committee later credited Soldini with the time lost during the rescue and he went on to become the overall winner of *Around Alone* and an international hero.

Three remarkable space-based systems: navigation, communication, and search and rescue made this rescue possible. The SARSAT system has been credited with saving thousands of lives since its inception. During one month alone, August 1999, 32 persons were rescued from ship disasters through use of the SARSAT system. But this system is vulnerable to storms in space. Storms on the Sun associated with solar disturbances emit radio noise in the 406 MHz frequency band that can swamp the emergency beacon detector on the GOES satellites. Fortunately the Sun was quiet during Isabelle's rescue.

The GPS, originally designed for the US military, has revolutionized many aspects of commerce including navigation and surveying and has been a boon to boaters and sportsmen. Fishermen mark their favorite fishing spots with GPS waypoints. As a runner, I use GPS to log my daily training distances and keep track of my speed. The GPS is now the basis of a line of navigation and rescue accessories used in automobiles and it is used in the trucking industry to monitor truck locations.

Geostationary communication satellites provide telephone, television, beeper, business data communication as well the email and fax services used by *Around Alone*. Most recently, a new fleet of low-Earth-orbit satellites is making cellular phone service available worldwide.

We take for granted the weather forecasts from geostationary weather satellites. Geostationary orbit is also the orbit of choice for military satellites including surveillance spacecraft that monitor facilities throughout the globe, navigation and communication satellites. Science and research satellites perform important tasks like observing movements in the Earth's tectonic plates to help forecast earthquakes, monitoring ocean temperatures and fish movement, keeping track of the Earth's ozone layer, and magnetospheric storms. Telescopes in orbit have revolutionized astronomy by making observations above the Earth's atmosphere.

More than 600 satellites are currently in orbit providing untold benefits to mankind. The monetary value of this space infrastructure must be close to $100 billion. Moreover, the space industry is growing exponentially with hundreds of new satellites appearing yearly. The majority of these are commercial.

Space technology has become integrated into the fabric of our lives in ways we would never have expected. But, like the oceans of the Earth, the new ocean of space is not without its hazards. The unmanned spacecraft that sail on the new ocean must also deal with adverse weather—storms in space.

On January 20, 1994, a magnetospheric storm knocked out two Canadian geostationary communication satellites, Anik E1 and Anik E2, and threatened a third satellite. Anik E1 was recovered quickly but Anik E2 required heroic engineering efforts before it could be returned to service several months later. These satellites cost about $225 million not including the $50 million or so for launch costs.

The Anik experience is not an isolated incident. Because satellites can't be brought back for examination, the exact cause of satellite failures is often difficult to determine. There is a reluctance to blame the space environment for satellite failures; however, it has been estimated that as many as 12 satellites have been lost as a result of space weather effects in the last 16 years.

Most commercial satellites are insured but the Canadian satellites, as for most government and military satellites, were not. The insurance industry is understandably concerned about increasing claims for launch vehicle and satellite failures. Space weather related failures have contributed to this growth in claims. Christopher T. W. Kunstadter of United States Aviation Underwriters, Inc. estimates that 'space weather has been suggested as a cause or contributor to over $500 million in insurance claims in the last four years'. He adds that 'numerous other cases of reduction in capability and loss of redundancy have been attributed to space weather' and that 'space weather is poorly understood and little appreciated for its effects by the space insurance community' (Kunstadter, 1999).

Catastrophic satellite failures make the headlines when service is disrupted; however, lesser satellite operational anomalies, commonly referred to as phantom commands, are much more numerous. Phantom commands include sudden and unplanned changes in the operating mode of a satellite. Usually these are detected and corrected automatically by commands sent from the Earth-based operations center. NOAA's National Geophysical Data Center in Boulder, Colorado, maintains a database of satellite operating anomalies based on information provided voluntarily by satellite operators. Statistical studies of the times and location in space of these anomalies indicate environmental conditions are the root cause. Figure 8.1 (color section) shows the location of satellite anomalies along the geostationary orbit compared with the MSM prediction of where the particle fluxes are high following a new magnetospheric storm. The next chapter, an interview with Joe Allen, documents some of the evidence linking satellite anomalies to the space environment.

How is it possible that a handful of miniscule particles like electrons flying around in space can cripple or even destroy a one ton satellite representing the very latest in electronic technology? The answer lies in the charge carried by the electrons. Our storms in space can, when they become sufficiently intense, immerse the satellites in a 'cloud' of hot electrons that charge the surface of the satellites to negative voltages. These voltages may become high enough at certain points on the surface of the satellite to produce sparks. Sparks are the arch-enemy of electronic circuits because they cause electrical transients that look like real electronic signals—phantom commands. Storms in space produce miniature lightning storms on the surface of the satellite. Just as in terrestrial thunderstorms, the lights sometimes go out.

There are postively charged protons and other ions along with the electrons in space, so why is there a build-up of negative charge on the satellites? The motion of the charged particles that brings them in contact with the satellite is the random thermal motion of any gas. Any particle in motion has kinetic energy. A particle's kinetic energy is proportional to the square of its speed and inversely proportional to its mass. Electrons and protons tend to have similar kinetic energies but, because the electrons have much lower mass, they have higher average speeds. The higher average speeds of the electrons bring them into contact with the surface of a spacecraft more frequently so there is a net buildup of negative charge on the surface. This charge buildup will continue to grow until the negative voltage on the surface repels the electrons. The voltage on the satellite is a measure of the average kinetic energy of the electrons in the space plasma. It may reach 10 kilovolts or higher in an intense storm.

Electrons with higher kinetic energy, those associated with the Van Allen radiation belt, can even bury themselves deep in the skin of a satellite and build up charge below the surface until there is an electrostatic breakdown of the surface material (deep dielectric discharge). Because of their potential to destroy a satellite, these electrons have earned the name 'killer electrons'.

Energetic protons from the radiation belts, the Sun and interplanetary medium can penetrate to the electronic circuitry and produce changes in the electronic state of memory chips by direct ionization of the semiconductor (single-event upsets). They also produce permanent and irreversible degradation of the solar arrays that power the spacecraft.

It might be expected that improved engineering techniques would mitigate space weather hazards to spacecraft as newer satellites are designed and launched. This has happened to some extent; however, several factors seem to be working in the opposite direction. First, micro-scale electronics have compressed greater numbers of components into smaller and smaller volumes, increasing the vulnerability of circuits to electrical transients. Second, it is now some 40 years since the dramatic discovery of the Van Allen radiation belt. A new generation of young engineers is less keenly aware of the hazards of space radiation. Examples of this were seen when the Hubble Space Telescope and more recently the Chandra X-ray Telescope both encountered problems due to energetic trapped radiation. Third, the increased demand for commercial satellites could exceed the number of experienced manufacturers and engineers. As space activity heats up and

competition drives down the cost of commercial satellites, less attention might be paid to satellite operational design safety. Newer
systems might be designed with inadequate thought to environmental
hazards.

Low-altitude, high-inclination satellites can be upset by space
storms for somewhat different reasons. Heating of the ionosphere by
the precipitation of energetic particles causes the ionosphere to swell
to higher altitudes. This can increase the atmospheric drag and alter
low-altitude satellite orbits. This problem applies particularly to satellites passing through the auroral zone and is most troublesome during
solar maximum when the ionosphere is additionally heated by
increased solar radiation. During one particularly severe storm, the US
military, which tracks satellites and space debris, reported a number of
lost satellites. It also found numerous new unidentified objects as the
computers struggled to reacquire and identify objects that had been
thrown from their nominal orbits by the storm. High-inclination orbit
satellites can also suffer periods of uncontrolled tumbling due to
torques introduced by strong magnetic fields in the auroral zone. These
satellites can also experience surface charging due to auroral electrons
(Maynard, 1995).

The hazards from storms in space are not limited to orbiting satellites. On the ground several types of systems are vulnerable. Perhaps
the most serious are the power grid blackouts caused by ionospheric
currents associated with intense auroras.

At 2:44 AM on the night of March 13, 1989 a giant magnetospheric
storm produced auroral-zone ionospheric currents that resulted in
mirror currents in long electric power lines that run along the length of
Quebec Province in Canada. These currents caused damage to AC
transformers that provide power to customers of Hydro-Quebec. The
entire Hydro-Quebec power grid was shut down. Six million persons in
Canada were without power for nine hours. Fortunately, the storm
came in the early-morning hours when electric demand was low.
During peak-load hours the outage could have cascaded to power pools
that serve the northeastern United States (Kappenman, 1993). In

addition to the lost revenue from the power outage, repair and replacement of the damaged transformers, and installation of protective equipment costs were in excess of $1 billion (Blais and Metsa, 1993).

The Hydro-Quebec case is the most severe on record but it is not an isolated incident. Sporadic blackouts and equipment damage have occurred in the USA and other countries as well. The most susceptible areas are those near the auroral zone. As with satellites, vulnerability to storms in space may be on the increase for power grids as well. In the USA, a move toward deregulation and increased competition will encourage electric energy suppliers to maximize the use of systems and minimize costs. There is also a trend toward interconnectedness of power grids to increase shared loads.

Currents from magnetospheric storms can affect long conductors used for communications such as transatlantic cables. Fiber optic cables are not immune. In fact, the first observation of a space storm influence on telecommunications was an induced voltage on a telegram line that was so large that operators discovered that telegraph messages could be sent without any batteries connected to the system (Lanzerotti, 1983).

Currents and voltages from magnetospheric storms can also affect long pipelines. Pipelines in moist soil experience corrosion due to metallurgical differences. This corrosion is exacerbated as currents flow through the junction between the pipe and the soil. Mitigation methods are used to overcome corrosion but engineers have become aware of the need to pay attention to geomagnetic storm-induced currents.

Even the manufacture of electronic microchips is not immune to the effects of storms in space. An integral part of this manufacturing process is the testing of each device on the assembly line. This testing requires the measurement of very low-level currents. Changes in the geomagnetic field can induce currents in the measuring process or power lines that invalidate the test being done. The result may be the rejection of an inordinate number of chips at that time; the quality control process becomes anomalous and the manufacturer loses money. The relationship with storms in space has been well established. In one study, 78% of peak-defect days were within one day of the peak geomagnetic activity for the six-week period (Pratt, 1993).

Ionospheric effects associated with storms in space cause problems for navigation and communication systems. A GPS receiver computes its position by radio signals from multiple satellites that carry precise information on the location of each satellite. The receiver also computes its distance from each satellite by measurement of the signal propagation delay time. This propagation time depends on the number of electrons in the ionosphere along the line of sight to each satellite. Changes in the number of ionospheric electrons due, for example, to a geomagnetic storm require corrections to the GPS position calculation. Several methods are available to reduce this error, however, rapid changes in the propagation delay time can cause the receiver to temporarily lose the satellite signal (Kleusberg, 1993). Rapid small-scale changes in ionospheric density, called scintillations, are hard to compensate for.

Other radio communication or navigation systems can be upset by storm-related modifications to the ionosphere. The period of the great storm of March 1989, which caused problems for electric power grids, was also marked with ionospheric disturbances that produced operational problems for the LORAN navigation system worldwide. At the same time ionospheric conditions were causing problems for GPS users, high-frequency radio communications were disrupted by the disturbed ionosphere so that ships at sea could not be notified that a storm in space was to blame for their navigational problems (Allen and Wilkinson, 1993). Military radio bands are frequently adversely affected by space storms. A crucial element of military logistics focuses on determining which radio frequency bands are available for communication during a storm.

Perhaps no other group is more keenly aware of the important role of the ionosphere in radio wave propagation than amateur radio operators. There are over 1 million amateur radio operators throughout the world. Hams provide a valuable service by providing communication services in times of natural and man-made disasters. They are interested in conditions affecting the ionosphere and are one of the largest group of subscribers to forecast services for solar-terrestrial conditions. Hams can take advantage of storms in space for improved communication. Increased electron densities in the ionosphere favor the reflection of radio waves at lower frequencies making long-range communication in certain wavelength bands possible by repeated reflection of the signals between the ionosphere and the Earth. The trick is to know which frequencies to use at any time. For this information, amateur radio operators rely on alerts and warnings of space storms issued by NOAA. The military faces the same problem of selecting the best radio frequencies for optimum communication. Several ionospheric models aid in the nowcasting and forecasting of ionospheric conditions.

We can clearly see that there is an urgent need to have a forecast capability for space storms. Commercial and military satellite operators, electric power companies, navigation system operators, amateur radio operators and many others all have a need to know when the storm in space is coming.

9 A Conversation with Joe Allen

When I met Joe H. Allen in his office on the morning of November 22, 1997, he was finishing an email message to an owner of racing pigeons who had written to ask Joe's advice on how to avoid losing pigeons. Joe explained to me that pigeons use magnetic sensors within their brains to help them find their way to the mark. Since a good pigeon can cost $10,000 or more and many in a flock may get lost during a race because of changes in the Earth's magnetic field, prediction of magnetospheric storms is a matter of vital concern.

Few busy scientists would have the concern, patience or interest to respond to such a request personally. Joe, however, is unique. In his recent capacity as Chief of the Solar-Terrestrial Physics Division of the NOAA National Geophysical Data Center and Director of World Data Center A for Solar-Terrestrial Physics, he has observed and collected information about the effects of the space environment for nearly 30 years. It was this great storehouse of accumulated knowledge that led me to ask him to do an interview. The following are some of the highlights of the conversation that took place that morning.

John: Joe—you were one of the first people to become aware of space environmental effects on human technological systems. What are some of the first things that you remember from the early days that triggered this association for you?

Joe: I can tell you the very first. It was one of those rare events where you know when it started. But, when you say I was one of the early ones—that's true in a way. But, really, I was one of the ones who talked about it. There were a number of people who were concerned about this who worked for aerospace companies and who worked in NASA, and they were aware of these things, but no one was doing much about it. The commercial people did not—and still don't—want anyone to know that their satellites have operational problems in space.

What happened—there was a UCLA graduate student named Wayne LeJeune.

John: About when was this, do you recall?

Joe: Yes. It was in August of 1972. This was an important date in our field.

John: Ah yes. There was a big space storm that month.

Joe: Not only was it a large storm, it was a phenomenally big solar energetic proton event. This was the granddaddy of them all and we are not going to have to worry about this happening very often. It happened then and another big episode happened in October of 1989 and it will happen again, but not frequently.

Wayne was working as a contractor for TRW while a graduate student at UCLA. They had sent him out here to the NOAA Forecast Center with some information about some geostationary telecommunication satellites that had severe operating problems. I don't know the details because they were classified. While he was here, he got a call from TRW to find out all that he could about what was happening with this storm because they were having problems. Later I heard that they lost at least one telecommunication satellite. Over at the NOAA Forecast Center someone suggested that he might want to talk to Joe Allen in the NOAA Data Center because he has been working with the auroral electrojet (AE, ionospheric current) and the AE index — high-latitude phenomena. Wayne LeJeune came over to talk to me. He had with him a diagram that showed a polar view with the Earth in the center — day on one side and night on the other — and plotted in a big ring around the Earth was the location of the telecommunication satellites' operational problems or failures. The problem area stretched from about an hour before midnight to the dawn meridian and most problems occurred about 0300 satellite local time.

I said, 'Oh Wayne, I can show you something that matches up with that!' I had this graph I had put together of the north polar region that showed the local times of the magnetic AE observatories that showed the most extreme magnetic deviations from hour to hour over a year of observations. Basically it showed the local times where the auroral electrojet was most intense most of the time. These local times matched the regions where the satellites had the most problems.

John: It peaked at 3:00 o'clock?

Joe: Yep, or 0330. His satellite anomalies laid almost perfectly over the same local time sector as the westward auroral electrojet, the one providing the AL index.

John: As far as you know this was the first time anyone had a clue that there was a connection between the satellite anomalies and geomagnetic storms?

Joe: Yes. It was so obvious that I couldn't believe that someone had not noticed this before. I gave a paper on this at the next Spring American Geophysical Union Meeting in Washington. Afterward someone from Aerospace Corporation took me aside and said 'We know about this.' I think it was one of those things where the satellite builders and operators were aware of it, but you kept running into people mainly on the operational side who did not know.

LeJeune went back to work at TRW and they started putting their satellites into plasma test chambers. The way I heard the story later from Wayne was that when they did this they started getting buildup of charge on the dielectric on the satellite surface and then there would be arcing between different parts of the surface and the satellite frame. This arcing could change the potential to which the internal components were grounded. This could look like a signal to the satellite so you could get phantom commands. An instrument that was on could turn off as if it had been commanded from the ground to turn off, or the other way around, all at an inappropriate time. So this was the whole family of phantom commands.

The phantom commands were distinguishable from another type of anomaly called 'single-event upsets', or SEUs. With single-event upsets a problem would occur once in an isolated event. We got to plotting the single-event upsets on the same kind of a plot and we found them occurring at virtually any time. Some single-event upsets were associated with bulk charging as opposed to surface charging.

John: How is bulk charging different from surface charging?

Joe: Surface charging is associated with lower-energy electrons that spray the surface of the satellite with charge. Bulk charging is caused by more energetic electrons that can penetrate more deeply into the skin of the spacecraft and build up a charge below the surface. This continues until the dielectric strength is exceeded and an arc occurs.

Very-high-energy protons and heavier ions, which pass right through the wall of a satellite, can penetrate directly into a chip and change the state of a transistor or burn through to form a destructive path. This isn't bulk charging but is usually the cause of a SEU or a permanent 'latch-up'.

John: Are there any other effects that can happen on a daily or even seasonal pattern?

Joe: Very-high-energy (relativistic) protons can permanently degrade the output of solar arrays and star trackers. Also, I once had a call from a satellite engineer with an aerospace company and he told me they were not seeing this midnight to dawn pattern, but rather they were seeing their phantom commands clustered at dawn and again at sunset. So I asked him if there was a seasonal pattern to the anomalies that they saw. He said 'No, why would I have that?' I said 'Because, when looking at the phantom commands, in addition to the daily effects we also see peaks in the surface charging at the equinoxes.' He said 'We'll look at it.' He came back to me later in the fall and said 'Well, that was great. It really pays to talk to someone who doesn't know what they're talking about.' (I thought that was a real backhanded compliment.) He said 'There's a good seasonal pattern to this.' These were geostationary communication satellites that had big dish antennas pointed continuously at the Earth. What he concluded was that sunlight-produced photocurrent caused by photoelectrons leaving the surface could neutralize the surface charge caused by the thermal plasma electrons arriving at the surface. When the satellites were at noon and midnight, the big dish antenna pointing at Earth would provide a large cross-sectional area to catch the Sun. But at dawn and dusk that cross-section of the antennas to the Sun would be minimized and the surface charging would be enhanced along with the phantom commands. The effect would be maximized at the equinoxes because, again, that was when the relative changes in the cross-section of the antennas to the Sun would be greatest.

Joe: You asked earlier about the bulk charging due to the higher-energy electrons. I don't think we began to appreciate the importance of this as a separate source of the effects of magnetospheric electrons on satellites until 1994.

John: What happened then?

Joe: The two Canadian communication satellites Anik E1 and Anik E2 failed within a few hours of each other. One was the eastern Anik, covering eastern Canada and the other was Canada's western Anik. When they both went out there were a number of people calling to find out what might have happened. There were some people who thought it might have been foul play or sabotage. I got calls from the Canadian Ministry of Defense. It was using contract services on these commercial satellites for communications purposes. They had to substitute ground radio communication for the satellite links. A lot has been published about what happened when the Aniks failed in terms of things like hotel and airline reservations and other commercial problems.

John: What really caused the failures?

Joe: I told them I did not believe in satellite anomalies without either a strong source of electrons or a solar proton event. So, I went first into the energetic electron data. There was an ordinary daily variation in the electron fluxes, nothing unusual, so I called over to talk to Dave Speich. I said, 'Dave, I just don't believe this is an environmental problem unless there is something unusual in the electrons.' He said 'Well, let me look at that.' He came back and said 'Oh my gosh, Joe, these levels are a thousand times higher than quiet background, even though they are varying in the regular diurnal pattern.'

John: You mean the peaks in the electron fluxes at certain local times were higher than they would normally be?

Joe: Yes. The amplitude of the high and low daily peaks was much higher than normal and it had been that way for a couple of weeks before the Aniks failed. These were higher-energy electrons so this was apparently a bulk charging induced failure. With this type of failure we know that the electrons can or must build up their concentration for some time before the subsurface charge is great enough to cause the arc that does the damage.

John: Let's recap. You have mentioned that there are three ways in which the space particles can affect satellites. There is the surface charging which tends to occur in the midnight to dawn region of the geostationary orbit. Then there is the deep dielectric or bulk charging, which apparently has no particular preferred time of occurrence, presumably because the charge takes a long time to build up. Finally there are the solar energetic protons, which can degrade the power output of solar panels.

Joe: Yes, but also these solar energetic particles (protons and heavier ions) can penetrate into electronic chips within the body of the spacecraft and cause a false signal or burn a path through the chip that is physically damaging.

John: Is this what is called a single-event upset?

Joe: It can be, yes. There is some confusion in the terminology. A phantom command is an operation performed by the spacecraft when it was not supposed to be happening or was not commanded from the ground. This is an anomaly. Single-event upsets are also unexpected circuit changes but they may not be operational changes. These are anomalies as well. On the other hand, the deterioration of solar panels during solar energetic particle storms is not generally called an anomaly. In operator terms this is a nonreversible degradation of the power output of the solar panel or solar cell array. This happens gradually all the time. You might have a ten-year design lifetime for a power panel but you can get a year's worth of degradation during a single solar energetic proton event or storm. Bulk-charging events are usually arcing and may duplicate phantom command effects or burn out a circuit element.

John: What energy protons are implicated in the damage?

Joe: There has been some debate about this. If the proton energy is too high, much above 10 MeV, they can penetrate all the way through the solar cell and not do as much damage, but generally the harder the spectrum the more severe the damage internally. Spacecraft optics as well as solar arrays can be affected. Limb sensors that find the edge of the Earth's atmosphere are often used to control the orientation. For example, GOES 8 and 9, three-axis stabilized spacecraft, use not only limb sensors but also star pattern sensors to help in maintaining the satellite's orientation toward Earth. During the proton event that started on the 4th of November 1997, Pat Macintosh sent out a forecast calling attention to region 8100 on the Sun a couple of days before it really flared. It produced an F9.4 X-ray intensity flare, which is big. Relativistic particles began arriving at Earth on the 5th of November. GOES 8 and 9 both malfunctioned because of the star trackers getting upset by scintillations from the high-energy protons striking the optics.

John: You mentioned Pat Macintosh. He worked as a forecaster at the NOAA Space Environment Center (SEC) didn't he?

Joe: Yes, he was the top forecaster at SEC. He retired the same year I did.

John: And he is still doing forecasts as a private operation?

Joe: Yes, he puts this out on email. It's a labor of love. Pat developed the active sector classification and the computer-based, neural network expert system forecast capability used by SEC.

John: Some time ago I heard Pat give a paper in which he said that the computer expert system couldn't beat the human forecaster. Is that still true?

Joe: That's right. He doesn't believe it's quite as good as he is. Now, I have heard Pat say that sometimes the computer can beat the forecasters but other times it doesn't.

John: To return to the space storms, was this November storm the one that damaged India's communication satellite?

Joe: No that was a month earlier—in October. I think the Indian satellite problem was associated with one of these low-level magnetic storms. These are storms that occur about two days after a coronal hole passes central meridian and the stream of high-speed solar wind sweeps by Earth. Look at this monthly plot of the flux of greater than 2 MeV electrons. Joe shuffled through a stack of pages and pulled out a plot. Notice that the flux of greater then 2 MeV electrons drops down for about two days then steps back up to higher than it was before the storm. That's characteristic of these recurrent low-level storms that happen quite frequently during solar minimum. Then some mechanism raises the level of high-energy electrons at geostationary altitude.

John: Each one of these storms results in an increased flux of these energetic electrons that can threaten the health of satellites?

Joe: That's right.

John: Where do you think these electrons are coming from?

Joe: They are probably not coming from the Sun with these energies. They may be entering the magnetosphere from the tail and then being accelerated in the magnetosphere.

John: That's interesting because we have just written a paper that presents evidence that they come in from the tail. I think that as they enter from the tail they are held in the outer region of the magnetosphere until the ring current starts to decay away. As this happens, the Earth's magnetic field returns to a more normal configuration and the energetic electron drift paths spiral inward to the geostationary orbit. They gain energy as they move inward. But tell me what finally happened to the Canadian satellites?

Joe: The power supply for the Anik E-2 pointing control system was severely damaged in February of 1994. The pointing control system on Anik E-1 was damaged at the same time but was recovered by switching to backup circuitry. It failed again on March 26, 1996. The backup circuitry for Anik E-2 had already been lost so it could not be used in February of 1994. They had to come up with an alternative scheme of controlling the pointing of the satellite using the onboard station keeping thrusters and fuel. They would sense the loss of signal from the ground as the satellite drifted out of its proper orientation and send a command from the ground to reorient the satellite using that system. Both of these satellites were early in their design lifetime when they failed.

John: What is the cost of a satellite like the Aniks?

Joe: A conservative estimate would be about $200 million for the bird itself. This does not include the cost of the launch, which would run well over $50 million.

John: Were there other times when the high-energy electron flux was high that satellites did not experience problems?

Joe: As I remember the same question was asked earlier. In 1995 there was a similar period of high electron fluxes but no damage. It appears that it is not just the peak or the magnitude of the electron flux but rather how long the high flux lasts. The duration of the high flux of electrons with sufficient energy to penetrate the satellite skin and build up sufficient charge is apparently what is important. It's a matter of the total cumulative fluence.

John: How do you keep track of or monitor these energetic electrons?

Joe: The geostationary GOES satellites operated by NOAA have detectors that measure the flux of electrons of energy greater than 2 MeV. They also have detectors for the solar energetic protons. The Los Alamos National Laboratory also flies energetic particle detectors on military spacecraft as well.

John: As I recall solar protons can produce background noise in the energetic electron detectors. This causes problems at times, does it not?

Joe: Yes, but usually some information can be obtained on the electrons.

John: Do you keep records on the frequency of satellite anomalies reported to you by various sources?

Joe: Yes. We started a satellite anomaly database in 1981 and we have collected event descriptions back to early 1971. At that time most of the satellites were geostationary satellites. For example, here is a plot of the anomalies in that database versus time. Generally the number of anomalies seems to follow the distribution of magnetic storms—that is until you get right here. This time corresponds to the launch of the NASA TDRS-1 (Tracking and Data Relay Satellite). TDRS-1 had thousands of anomalies. It was what one fellow described as a flying cosmic ray detector masquerading as a memory chip. It used banks of chips. They fixed the problem by changing that bad memory chip on the next satellite. They didn't want to make the change because of the cost. They had one TDRS in orbit and the second one sitting waiting to be launched, and we had our meeting here in October of '85. There was a fellow I was corresponding with at White Sands in 1984/86, Don Wilson, who was doing check-sum miss-compares recorded with a little computer he bought at K-MART. The check-sum miss-compares were matching up so perfectly with magnetic storm disturbed periods that we were convinced that they had a component that was bad. After our workshop they made the change and the problem almost disappeared.

John: You have continued to build the anomaly database over the years. Do the satellite operators always report their anomalies? I understand there is an increasing reluctance to report problems.

Joe: Unfortunately they don't really like to report anomalies. Everyone has their reasons, usually commercial, for not being eager to share their problems. As a result there is very spotty and incomplete reporting. That's why I have started this new email newsletter. This is sort of a periodic update on recent events that goes to a small mailing list. It is also a request for information to see if we can change this trend of reluctance to report.

John: We have used the National Geophysical Data Center anomaly database ourselves and it has been a tremendously valuable resource. Let's move on to another subject. I have wondered for some time about the effects of the radiation belt electrons and protons of radiation degradation on solar panels. This is not usually discussed as a major factor. Is it considered to be important?

Joe: Housekeeping observations on the solar panel power output are made by the satellite operators on a routine basis. I have personally seen these for the GOES satellites. They measure the power out into the fully loaded system. What you see when you look at these plots is a daily oscillation, and superimposed on this is a gradual downward slope and a yearly cycle. This slope can be compared with the design lifetime. You can see that the effective lifetime of the satellite is determined, in part, by the degradation caused by radiation during quiet times. I don't think this has been a big limiting factor, however. Other systems onboard usually die before low power gets to be a problem, except when you have something like the September 1989 storm that took a year off the projected lifetime of some satellites—and it didn't go back up. That was followed by the October storm that took three to five years of lifetime.

John: Three to five years of useful lifetime! This was for the NOAA/GOES satellites?

Joe: Yes. I know that there were military satellites that had problems at the same time but probably the radiation hardening specifications then for the military spacecraft were more rigid than for the commercial and government satellites.

John: Joe, I hear it said quite often that these storm-induced satellite anomalies and degradation problems are basically spacecraft design problems that can be eliminated with proper engineering, and we don't need to worry about trying to monitor or forecast space weather. I would like to have your thoughts with the long-range view on this question.

Joe: They can be engineered around when the engineering design and construction are supported at the right financial level, but there are always financial constraints. That means all the right work never quite gets done. It's like the statement you see at the end of nearly every PhD Thesis: 'This is an interesting subject that should be studied in greater depth.'

There was a meeting on July 4, 1985, at the Los Alamos National Lab to discuss anomalous satellite effects. I gave a talk on magnetic storms and magnetic indices. At the end of my talk a young officer from the Albuquerque Weapons Lab stood up and said 'What you're telling us is that magnetic storms cause surface charging events and we've already solved that problem —we've engineered around it.' An old fellow in the audience who was one of the senior aerospace corporation scientists answered and said 'Son, you're probably not cleared at a high enough level to know that you have already lost some satellites recently to surface charging and, furthermore, the person who approved the reduction in hardening of those satellites should be shot.'

One of the early failures was the '82 failure of one of the GOES satellites. I think it was GOES 4. I'm not sure, but it happened in the latter part of the calendar year, around Thanksgiving. The satellite kind of limped along for a few months and they pushed it along, trying to keep it going. The conclusion was that there was some engineering failure. I got to looking at it and it was a case were there had been a substantial amount of solar activity: a high level of electrons—not energetic electrons, although we were aware at that time of bulk charging—and there was a solar proton event. All of this happened before or concurrently with the failure of GOES. I said, how could that be an engineering failure with all of this going on unless there is something in the engineering that is unable to sustain itself during the disturbance? I guess you can say everything is an engineering failure—if you want to call it that. The environment had to be playing a role in this. Well, I started talking about this back at NOAA, at Satellite Headquarters, and at the Data Center. I tried to show why keeping solar terrestrial data at the Data Center is worth spending money on. This has paid my salary and the salaries of my people. The Air Force people were convinced we had something, but it was just the NOAA satellite people that didn't much like it. NOAA contracted out its satellite purchasing and satellite systems to NASA's Goddard Space Flight Center (GSFC). Harry Farthing from GSFC had chaired the Tiger Team that studied that GOES failure. Their original conclusion for some reason was that it was an engineering failure. Harry published a NASA internal report in which he did an analysis of the phantom commands on a couple of GOES satellites. He took the start time of a substorm, T_0, at the midnight meridian as determined by my AE index, and the time of a phantom command that seemed to be related to that substorm, T_f, and he computed the time difference in these two times, delta-T. He did this for a number of events. What the data showed was a beautiful linear progression with greater delta-Ts for greater distances of the satellite around from midnight. He plotted the slope of the line formed by these delta-Ts and the satellite distance from midnight, and found that this corresponded to the longitudinal drift time of electrons. The electrons could be injected onto the geostationary orbit from the magnetotail near midnight and, since they drift faster than the satellite moves, they would catch up with the satellite and engulf it in a cloud of plasma and contribute to surface charging. The farther the satellite was from the point of electron injection—near midnight—the longer it would take for the phantom commands to occur. This was all in an internal NASA publication. Later in October 1985, after Harry saw our particle data in my talk, he said that clearly the failure was a space environment response.

John: It seems that the evidence is almost irrefutable that the environment is damaging spacecraft.

Joe: In the late 1980s we had a meeting of the team at NOAA that was to discuss the next generation of instruments that was to be flown on the GOES satellites. I was invited to talk to them and this fellow who was head of procurement operations at NOAA introduced me and said 'Well, Joe Allen is here again and he is going to try and convince us that the Sun has something to do with how our satellites operate.' He was dead serious. So I shuffled my vugraphs and said 'I'll go ahead and show you not just the effects on the GOES satellites but for a number of other satellites as well.' I started out with this list of satellite anomalies from our database. Here were the particle records along with the list or records of failures of all of the different satellites. On the list were the failures that I said were almost certainly due to the environment. The committee was convinced. Years later, after the ANIK failures, I met an English fellow who had been looking at a series of occurrences of anomalies on what he called DRA-Deltas, a NATO satellite series. He couldn't say which satellites they were but he had the anomaly data that he could talk about openly. He said 'Joe, you ought to say we can be absolutely certain,' because his anomalies followed the greater than 2 MeV electron peaks perfectly and just clustered around the failure times of the ANIK satellites. They really were beautifully matched up with the anomalies on the other satellites.

John: We know that the energetic electrons show a strong diurnal enhancement. That is, we know that their fluxes peak around noon. I am a little bothered about the fact that the bulk-charging anomalies can occur anywhere in local time on average. Is this perhaps because it takes a while, maybe more than 24 hours for the charge embedded in the satellite skin to build up to dangerous levels? It must really be a fluence effect and not an instantaneous flux effect. Maybe that's the answer.

Joe: I think it is. It's maybe a little bit like Russian roulette. Suppose you're charging a portion of the spacecraft and then you have an arc. Where is that arc going to go? It could go to a part of the spacecraft that is perfectly safe, or it might not.

John: Joe, so far we've discussed spacecraft effects associated with magnetic storms. Lets talk about effects down here on Earth. When I came in you were working on a letter to someone who races homing pigeons. What's the story there?

Joe: This dates from a message from Wade Wilde, a pigeon racer in the San Diego area, to Les Morris. Les is on my distribution list and he inquired about what I remembered about pigeon racing in the past. I referred him to Joann Joselyn over at the NOAA Space Environment Center because they have some pigeon racing customers who pay to receive reports of magnetic storms.

John: Yes, I remember hearing that.

Joe: Well, Les apparently forwarded my reply to Wade who sent some more questions back to me. I responded to him directly; summarizing: what's a magnetic storm, how are magnetic storms measured, what's a K index, Kp index, that sort of thing. This email I have this morning is his response to that, with a couple more questions. That is typical of the sort of thing we have gotten into.

John: Can you explain what the exact problem is with the homing pigeons?

Joe: Apparently there are two racing seasons a year. One happens— and I may have this reversed—in the Spring when they race the older and more experienced birds. And then, because of the hatching and training, in the Fall they race the new birds. The birds survive to get to the first category by just being successful in the Fall. They have flights of birds on some days that will be a bust. A bust will be when they lose 80 to 90% of the birds.

John: They don't arrive where they are supposed to go because of magnetic storms?

Joe: Yes. We used to make jokes about this. Maybe the hawks and peregrine falcons had little magnetometers onboard and they knew that the pigeons would be better prey whenever there was a magnetic storm in progress. You could just see the falcons going in wearing little leather helmets picking off the homing pigeons. But whatever it is, there are times when a flight of racing pigeons will just be decimated. One fellow who was studying this found that magnetic storms occurring during these racing periods would serve to disorient the pigeons. The disorientation seemed worse during electrically stormy conditions when there were clouds and the Sun was blocked from view. The primary navigation used by the pigeons is supposed to be Sun angle. I have heard of experiments at Iowa State where pigeons are raised in controlled day/night enclosures where they run the lights on 24-hour cycles but they move it ahead a little bit. Gradually they time-shifted the birds up to six hours from the local natural day/night sequence. Then they would release a flock of birds that had been subjected to this environmental alteration, brainwashed birds if you will. This shifts the angle of take-off to go to their home destination and it shifts it by the hour-angle between where the apparent Sun is and the time of day that they have been conditioned to inside the enclosure. This is taken to mean that the main method of navigation is Sun angle as well as some way of imprinting where they are and where they live and how to get there. However, if you rule out the use of the Sun angle, and maybe even if they have the use of the Sun angle but are nervous, maybe beginners, the occurrence of a secondary effect like a magnetic storm can cause them to become disoriented and lost. I understood that the reason they were contracting with NOAA Space Environment Center was so that they would not race birds on magnetically disturbed days. Of course the best forecast you can have on predicting magnetic storms is when you are having 27-day recurrent storms on the declining side of the sunspot cycle. What I pointed out to them is that the preponderance of magnetic storms is in the weeks around equinox, in the Spring and Fall. That happens to be prime racing time for them. So they are racing their birds in probably the worst time because of the more frequent occurrence of magnetic storms then.

John: Why are there more magnetic storms during the equinoxes?

Joe: According to the Russell–McPherron theory the equinoxes are the time when there exists the maximum cross-section of magnetospheric area exposed to the solar wind. This is because the tilt of the dipole axis relative to the direction to the Sun is minimized at this time. The energy available for transfer between the solar wind and the magnetosphere through magnetic merging is the highest.

John: I noticed that in the email from the racing pigeon owner he asked if there are forecasts of the Ap magnetic index. Are there?

Joe: At the present time the Air Force produces a 1-hour, real-time index that is similar to Kp in that it is an indicator of geomagnetic activity level. This is generated by a smaller number of magnetic observatories than the actual Kp index. We performed a study that showed that Kp and Ap, when converted to a 28-step index are very well correlated: a correlation coefficient of about 0.9.

John: Is this available to customers outside the Air Force?

Joe: Yes. NOAA has it on-line.

John: Let me change the subject slightly. You have spent a good portion of your life in the collection and preparation of these various magnetic indices that provide information on the magnetic storms that are responsible for these effects we have been discussing. What do you see as the contribution that these indices will make in our future ability to mitigate the hazards of space weather? Are we going to come to the point where we will be able to know when there is going to be an anomaly on a spacecraft and if so prevent the damage or loss of a several hundred million-dollar spacecraft?

Joe: I think that the conditions that contribute most to the damage of one satellite and not others at the same time and same altitude are more complex than can be inferred from an index alone.

   For a couple of years I looked at three-hour intervals of Kp of 0 on up to Kp of 9 and the number of anomalies seen at the same time. What I found was that when Kp was zero, there were no satellite anomalies reported in our database. When Kp was 1 or 2 there were very few. For Kp of 3 (and 3 is not all that unusual — I don't think you are getting into a big magnetic storm until a Kp of 5), the rate of anomalies was roughly double what you would have if you had a uniform distribution. And for every level of magnetic activity above 3 that held true. For each higher step in Kp you had about double the number of anomalies that you would have had above a uniform distribution. So this was the basic result. As soon as you got to Kp of 3 and above, you started seeing an increased number of anomalies. Now this doesn't tell you which satellite or where. It doesn't say a satellite is most susceptible in the midnight to dawn sector or, like that self-shadowing satellite we talked about, at dawn and dusk. There are just so many different things that are important. Remember that Japanese satellite we talked about that had died in February of 1994. When we plotted the location of that failure on the electron flux plots from your Magnetospheric Specification Model we found that the problem started right at the maximum gradient in the electron fluxes. Also the satellite had been subjected to a high fluence of energetic electrons for several days prior to that. And then the flare occurred and a big proton release came from the Sun with the solar energetic protons getting to the magnetosphere. So that satellite experienced the unfortunate concurrence of a high fluence of energetic electrons for several days, then the energetic proton event, and finally being at the position in its orbit of the maximum gradient as determined by your specification model. And bang, it went.

John: It had a lot of strikes against it.

Joe: So one index isn't enough. Although the index of magnetic activity would have been high for that three-hour interval, it had to be the rest of those conditions coming together.

John: What you are saying is that we need models that will give us more detailed information than we can obtain from an index.

Joe: Yes. I think that the models will reflect magnetic activity and the interplanetary conditions. That little gateway that lets solar energy into the magnetosphere is important. I think that models driven by data that somehow incorporate the indices and the real world are the only way I see to go. I think in the space weather business there may be too much emphasis on forecasting. I think nowcasting (instantaneous reporting of conditions) has a role to play as well. The models can tell you what is going on at locations where you don't have data and they can do it in real-time, nowcasting. You've got to have good models for that too. I think this may be as important as forecasting. Then there's the retrospective look back that can say historically what happened. Because maybe we didn't predict what happened. Or maybe we predicted something and nothing happened. There are so many aspects to prediction that to only sell space weather based on 'we will predict and we will do it successfully' alarms me.

John: So you think the role of space weather should also be nowcasting as well as retrospective analysis capability.

Joe: Yes, providing a framework into which you can organize your retrospective analysis. I think all three of these are important.

John: That's a very good point, Joe. Thanks very much for your time.

10 Manned Exploration and Space Weather Hazards

> By the time Claggett and Linley reached their rover and turned it around,
> they no longer bothered with their dosimeters, because once the reading
> passed the 1000 REM mark, any further data were irrelevant. They were
> in trouble and they knew it, but they did have a chance if they did
> everything right.

James A. Michener, *Space*, Random House 1982

Space, the great novel about the Apollo program, dramatically illustrates a real hazard for manned space flight. Michener adds a seventh mission to the Apollo lunar landing series; a mission to the back side of the Moon. During 'Apollo 18', a giant solar flare erupts on the Sun that engulfs astronauts with a lethal dose of solar energetic particles. While this episode is fiction, it is based on a plausible scenario. The Sun, on rare occasions, generates flares and coronal mass ejections capable of depositing at the orbit of Earth fluxes of high-energy protons that can kill an unprotected astronaut.

On September 29 and 30, 1989, a burst of very energetic solar protons associated with a large flare on the west *limb* of the Sun engulfed the Earth. Ground-based neutron monitors indicated that this was the most intense flux of energetic particles to reach the Earth since the start of the space age (Allen and Wilkinson, 1993). Estimates indicate that the radiation hazard for unshielded astronauts on the surface of the Moon or on a deep space mission to Mars would have been significant (Lett, Atwell, and Golightly, 1990). Fortunately, no such solar storms occurred during the real Apollo flights. NASA was aware of the potential for such an occurrence and planned for continual monitoring of the Sun during each mission.

Even high-altitude aircraft passengers are not immune. Dosimeters aboard the *Concorde* show that passengers and crew received a radiation dose equivalent to that of a chest X-ray during this storm (Allen and Wilkinson, 1993).

For astronauts there is still some good news. While a space suit does not afford much protection, a spacecraft can provide substantial radiation shielding. The exact amount of protection provided by the interior of a spacecraft depends on a number of factors: the amount of surrounding material to absorb the radiation, the energy spectrum of the charged particles in the storm, and the type of biological effects being considered. For the September 29–30, 1989, solar proton event it has been estimated that transferring from an *EVA* to the interior of a spacecraft with a shielding of 2 grams per square centimeter equivalent of aluminum would reduce the dose between about 1.4 and 4.2 times (Lett, Atwell, and Golightly, 1990).

The bad news: the extremely high kinetic energy of the particles means that they travel at close to the speed of light. Light takes only about 8 minutes to reach the Earth from the Sun; therefore solar energetic particles can arrive at Earth in less than an hour. This allows for very little warning. While we can observe disturbed regions on the Sun as they grow in intensity, there is no forecast model yet that will tell us exactly when a given region will erupt violently and which of these will spawn solar energetic particles that will reach the Earth.

The Earth's giant magnetic field serves to deflect solar energetic particles away from the Earth—except in the regions within about 30 degrees of the magnetic poles where the charged particles can slide down the magnetic field lines and easily reach the upper atmosphere. Most of the US space shuttle flights take place on low-inclination orbits and all are at low altitudes, safely below the reaches of these hazardous particles. The International Space Station (ISS), however, will be in a 51.6° (geographic) inclination orbit to accommodate the launch facilities of our Russian partners. Because the magnetic dipole axis of the Earth is tilted relative to the geographic axis, this orbital inclination will take the ISS slightly above 60° geomagnetic. Therefore, during intense solar storms, significant fluxes of solar energetic particles can reach the latitude of the ISS at certain longitudes. A great deal of the assembly of the ISS requires extravehicular activity where an astronaut has only a space suit for radiation protection. Studies indicate that even at these high latitudes the Earth's magnetic field provides sufficient protection that EVA astronauts probably will not receive a lethal dose of solar energetic particles from the largest storm (Committee on Solar and Space Physics, Space Studies Board, National Academy of Science, 2000). However, care must be taken to insure adequate warning time so that EVA activity can be terminated to allow astronauts to return to the relative safety of the Space Station. Meanwhile, scientists are struggling to find new ways to predict when the Sun will erupt with a barrage of energetic particles.

There is another radiation hazard associated with the higher inclination orbit chosen for the International Space Station, the intense trapped radiation of the outer Van Allen belt. The Apollo astronauts avoided serious exposure, on the way to the Moon, by passing through the Van Allen belts quickly. Furthermore they only had to make two passes through the belts. The space shuttle, on low-inclination flights, is well below the radiation belts. But at higher latitudes there are regions where the radiation belts extend to shuttle altitudes. The Earth's magnetic dipole is not only tilted relative to the Earth's rotation axis but it is also slightly offset from the center of the Earth. Because of this, there is a region over the south Atlantic Ocean where the outer belt presses against the upper atmosphere and reaches manned spacecraft altitudes.

The ISS orbit also grazes the auroral zone where energetic particles are continually leaking from the radiation belts. Unfortunately, because these regions are so close to the atmosphere, reliable models for the radiation dose rates don't yet exist. Furthermore, it is difficult to estimate how high they might become during a space storm. The best determinations of what doses to expect come from observations on manned missions.

A team of scientists at NASA's Johnson Space Center, the Space Radiation Analysis Group (SRAG), keeps tabs on the radiation environment of the astronauts. They make every possible effort to insure that the radiation exposure received by the astronauts is as low as reasonably possible. To accomplish this they use data from past manned flights, from unmanned real-time monitor satellites, from NOAA's Space Environment Center and other space weather centers, and from models. The job of SRAG will become more difficult in the coming years as EVAs used in space construction become more frequent.

Finally, for extended manned flights to other planets such as Mars, astronauts must contend with still another component of space radiation, galactic cosmic rays. GCRs represent the highest-energy space particles. Fortunately their flux is quite low compared with the solar energetic particles. They are protons and other bare nuclei of atoms accelerated to very high energies somewhere in the Milky Way Galaxy or beyond, by processes we still do not fully understand. The flux of GCRs capable of reaching the inner solar system varies with the sunspot cycle. Because an active Sun enhances the solar wind, which in turn acts as a barrier to galactic cosmic rays, the GCR flux is highest during solar minimum.

The highest-energy GCRs reach energies far higher than people can generate in laboratory accelerators. In fact they are our only means of studying elementary particle interactions at extreme energies. Upon entering the atmosphere, galactic cosmic rays undergo nuclear reactions with the nuclei of atmospheric atoms and produce secondary particles that continue interacting with still more atmospheric atoms to form a cascade. A single high-energy cosmic ray interacting with atoms of the atmosphere can generate a shower of secondary energetic particles and electromagnetic radiation that spreads out over a large area at the Earth's surface. Hundreds of cosmic ray secondary particles with far lower energies and less destructive power than their progenitors pass through our bodies each minute.

In space, individual cosmic rays can produce a path of ionizing destruction. While the radiation dose from GCRs for low-altitude, limited-duration manned flights is not a serious hazard, the situation is different for long-duration manned flights away from the protective magnetic shield of the Earth. GCRs are sufficiently penetrating that it is difficult to imagine an adequate, practical radiation shield. Except for the modest variations that occur over a solar cycle and temporary decreases in intensity during CMEs and other solar wind enhancements, galactic cosmic rays are an ever-present, low-level but potent space radiation hazard.

The accumulated radiation dose from galactic cosmic radiation for a round-trip to Mars lasting several years would be near the recommended limit for astronauts (Lett, Atwell and Golightly, 1990). This limit is based on the radiation dose expected to produce an increased risk of cancer of 3% over the astronaut's lifetime. There is also the possibility of one or more solar energetic particle storms during the trip. The surest way to mitigate these hazards would be to shorten the round trip time. A new generation of advanced, plasma rocket engines that could significantly cut the one-way travel time to Mars offers promise in this direction.

The exposure to galactic cosmic rays and the risk from a storm of solar energetic particles are just two of many hazards that astronauts will face on a voyage to Mars and back. As for myself, I prefer the protective cocoon of Earth's marvelous atmosphere and magnetosphere. Still, I admire those who willingly take the risks.

11 The Future of Space Weather Forecasting

Everyone talks about the weather but nobody does anything about it.

Colloquialism

Like terrestrial weather, space weather involves forces of nature so enormous that it is impossible to imagine human intervention at our present state of technological advancement.

We can't control space weather, but by trying to understand and eventually forecast storms in space we can hope to minimize their impact.

Until about 10 years ago space weather was the domain of the esoteric field of magnetospheric physics. As such, space plasma physicists studied it as a pure science. It had been known since the beginning of the era of Earth satellites that the space environment could interfere with the operation of spacecraft. Conferences on spacecraft charging and radiation hazards in space were held in the early sixties; however, it was not generally appreciated by most space scientists how much trouble storms in space could cause. Satellite operators, from commercial to the military, were understandably reluctant to discuss spacecraft problems, or even to share data on anomalies. As we have seen in Joe Allen's stories, there was a tendency to assume that the problems were not related to the space environment. Gradually, the body of physicists studying the Sun, solar wind, and magnetosphere began to appreciate that there were important practical reasons for understanding and modeling the space environment.

The hazards of storms in space are now sufficiently widely recognized that serious activity is under way to address the practical applications of our understanding of the space environment and to focus research on those areas not yet well understood. Under the leadership of the National Science Foundation (NSF), relevant government agencies including NASA, the Departments of Commerce, Defense,

Energy, and Interior, have joined forces to establish the National Space Weather Program. The overarching, stated goal of the NSWP is: '..... to achieve an active, synergistic, interagency system to provide timely, accurate, and reliable space environment observations, specifications and forecasts within the next ten years'. The NSWP envisions an infrastructure for space weather similar to the National Weather Service. Toward this vision, the NSWP has developed an implementation plan that includes research, and the development of models and products. The NSWP has the support of many scientists throughout academia, industry and government, and it is the focus of targeted research programs funded by the National Science Foundation.

One such NSF program is the Geospace Environmental Modeling (GEM) program. The stated purpose of GEM is to support basic research into the dynamical and structural properties of the magnetosphere. One major task of GEM scientists is the construction of a Geospace General Circulation Model. The GGCM is patterned after the atmospheric General Circulation Model that has become a major research tool of the atmospheric and climate modeling research community. Each summer about 150 space physicists and graduate students meet in Snowmass, Colorado, for a workshop to press the task of building the GGCM and related tasks aimed at understanding the magnetosphere. The work of building models that can eventually forecast space weather has become a labor of love for these scientists who represent about 10% of the larger community of magnetospheric physicists. The magnetospheric physics section of the American Geophysical Union has about 1,500 members most of whom are involved in one way or another with research relevant to space weather.

This book has focused on the magnetospheric aspects of space storms but the areas of atmospheric, ionospheric, solar and interplanetary research are equally important. NSF programs equivalent to GEM exist for these areas as well. These programs are CEDAR for the atmospheric and ionospheric research fields and SHINE (Solar, Heliospheric, Interplanetary Environment) for the solar and interplanetary medium areas. Model development is a vital research tool here as well.

The fundamental task of understanding all the processes responsible for storms in space remains a challenge for this small army of scientists.

NOAA is the main US government agency charged with making space weather information available to the public. This is accomplished at the Space Environment Center in Boulder, Colorado. At SEC, NOAA maintains a forecast center with a staff that watches the Sun, satellite and ground-based data 24 hours a day. They issue regular space weather advisories with alerts, watches, and warnings for space storms. These SEC space weather advisories can be received by fax, the Internet, and even by pager. The NOAA/SEC web page can be accessed on the Internet at www.sec.noaa.gov. This web page contains information on the latest space weather conditions. Included are the latest images of the solar disk, NOAA satellite data on solar X-rays and energetic particles, the Kp index as computed in real-time from ground-based magnetometers, a forecast of Kp based on solar wind data from the ACE spacecraft, and even the output of the Magnetospheric Specification Model.

At this time the SEC alerts and warnings are based mainly on the experience of highly trained forecasters. There have been efforts to train computer codes, called expert systems, to predict when an active region on the Sun will erupt and create a storm in space. So far, however, the human forecasters can usually beat the computers. Progress in building quantitative models that can forecast storms has been slow. Models are available that can track a shock wave that moves outward from the Sun through the interplanetary medium. When given adequate information about the conditions at the Sun, these models can predict when the shock wave will hit the Earth within several hours. Unfortunately they often forecast the arrival of the shock too late.

A great deal of effort has been focused on models of the magnetosphere. The Magnetospheric Specification Model discussed in Chapter 7 is the first full computational magnetospheric model to be used in a space weather operational setting in the United States. It has been used at the Air Force 55th Space Weather Squadron and at NOAA's Space

Environment Center. MSM output is also used by corporations to provide specialized products for commercial customers. However, the MSM is driven by data from satellites and also from ground-based magnetic observatories. This reliance on ground data means that the MSM is not capable of a true forecast. Instead it can provide a look at conditions throughout the magnetosphere about an hour after the conditions have established themselves. This does not mean that the MSM is useless as a forecast tool, however. Just as with terrestrial weather, space weather is a spatial as well as temporal phenomenon. Satellites move through regions of space just as they move through time. Knowing where the 'hot spots' are in space along with where various satellites are headed allows satellite operators to know when their birds will enter hazardous regions.

Nonetheless, the real goal is a true forecast model, a model that can tell us what will happen in the magnetosphere before it happens. This requires two things: first, a highly accurate model that can run on the computers faster that real-time; and second, input data that drives the model must be available to the model ahead of real-time at the Earth. The latter is possible if we take our data upstream in the solar wind or at the Sun. The solar wind reaches the Earth a little less than an hour after it passes the ACE spacecraft. On the other hand, the telemetry from ACE reaches us at the speed of light. The Magnetospheric Specification and Forecast Model, an advanced version of the MSM, finesses this problem by using special computer algorithms called neural networks trained to forecast the required ground-based input data one hour ahead of real-time. The MSFM has been prepared for use by the US Air Force. An advanced version of the MSFM eliminates the reliance on ground-based data altogether and runs on upstream solar wind data alone, affording an additional hour or so of forecast capability.

Models like the MSM have advantages and drawbacks. These first-generation models have the advantage that they are fast and can run on small computers. However, they are not fully self-consistent and are therefore less accurate and less complete in the suite of output data they provide. There are a number of more detailed and accurate

MHD-hybrid models under development, but they are computer intensive and slow. It is difficult to say when the first of these codes will become operational.

Still, the capability to forecast the onset of a storm a few hours ahead does not satisfy all of our requirements. The US military would like to be able to forecast storms days or even weeks in advance. This pushes the problem back to forecasting solar weather. It also entails more accurate models of solar wind propagation than now exist. This is where the work of SHINE, the NSF program that addresses solar and interplanetary forecasting, will play an important role. An intermediate solution would be solar-wind-monitoring spacecraft stationed closer to the Sun.

An additional ingenious project involves placing one or more solar-orbiting spacecraft so they can view the Earth–Sun line from the side. By watching changes in the radio signals from distant stars as the solar wind passes between the star and the spacecraft it should be possible to determine when CMEs are headed for Earth. When two spacecraft on either side of the Earth are used, this project, called STEREO (Solar Terrestrial Relations Observatory), could yield valuable data for the forecast of space weather.

The Van Allen radiation belts pose an additional challenge not addressed by present models. Late in most magnetospheric storms there appears a fresh supply of very-high-energy electrons that pose a significant threat to satellites. These electrons may even alter the chemistry of important atmospheric constituents in ways we are just beginning to discover. The exact source of these ubiquitous and dangerous high-energy trapped particles still eludes us. We have statistical models that can specify average conditions and scientific models that can explain some characteristics, but forty years after their discovery the radiation belts are still perhaps the least understood aspect of the magnetosphere. The Holy Grail of space physics is an accurate model of the energetic Van Allen radiation belt particles.

It seems clear that the forecasting of storms in space will require greater reliance on models. These models are not yet ready for use, in some cases because we don't fully understand the processes involved. However, our understanding of the magnetosphere, solar wind and Sun is growing as we develop the models.

To date, emphasis has been placed largely on models of individual segments of the space environment such as the magnetosphere, ionosphere, or solar wind, however, the future will probably involve a single model which will cover the whole of space from the Sun to the Earth. A group headed by the University of Michigan is already at work on a 'Comprehensive Space Environment Model' (CSEM). This particularly challenging approach requires a numerical model capable of functioning over a wide range of physical phenomena with vastly different scales of distances and times. The computational approach required by the CSEM breaks new ground in computer programming technology. It is designed to take unique advantage of the new generation of parallel processor, high-speed computers. Model building (software) and computer technology (hardware) must go hand-in-hand. Without sophisticated models such as the CSEM new computer hardware would never reach its full potential.

An important aspect of model building is visualization of the vast quantity of data generated by space environment models. We learn from the models we build but only when we can extract the information they provide. Unless the model output can be converted to images that can quickly be assimilated by the human senses and understood, the model is useless. This requires the exploration of new frontiers in computer graphics. We can visualize computer-generated pictures of giant prominences emerging from the Sun, surging through space, energizing the magnetosphere, generating the northern lights, intensifying the ionosphere and even heating the atmosphere.

Model building also requires a steady flow of new scientific data about the space environment. We must have data upon which to base the models and with which to test the models. Ultimately, forecast models are driven by data, just as is the case for terrestrial weather forecast models. This requires a continuing investment in new spacecraft to provide new data of ever increasing sophistication and resolution as well as spacecraft to monitor and provide real-time data over long periods of time.

Space weather forecasting today has been compared with the state of terrestrial weather forecasting several decades ago. As we have seen, the two fields have many things in common. Space scientists have attempted to learn from the experiences of terrestrial weather forecasters in order to telescope progress in space weather modeling and forecasting. We can look forward to the day when most television weather forecasts will routinely include a segment on space weather.

Ultimately, the pace of development of our understanding and capability to forecast storms in space will depend on the support of the public and private institutions. This support can only come through an awareness of the importance of space weather. It is my hope that this book can contribute in some way to building that support.

Mathematical Appendix
A Closer Look

Some concepts used in the discussion of space weather require mathematical definitions. These definitions allow the physical processes associated with storms in space to be described with greater precision. By making these mathematical explanations available in the appendix the meaning and appreciation of space weather may be enriched for the interested reader.

I ENERGY

Perhaps the most important physical concept in all of nature is energy. Energy is the currency of the physical world. Instead of 'follow the money', we might say, 'follow the energy'. Many important processes in nature are related to the transformation of energy from one form to another. Energy helps us keep track of what is happening in a physical process. Let's have a look at some of the different types of energy.

Moving objects have energy by virtue of their motion. The motional energy of an object is called 'kinetic' energy. The kinetic energy w of a single particle (moving slowly compared with the speed of light) is proportional to its mass m and to the square of its velocity V:

$$w = \tfrac{1}{2}mV^2.$$

Velocity is a concept like speed except that the direction must also be specified (velocity is a *vector* quantity). The speedometer in a car measures speed. 30 kilometers per hour is a speed. 30 kilometers per hour northwest is a velocity. Mass can be thought of as an object's resistance to a change in velocity produced by some force acting on the object.

Space weather calculations are often concerned with the properties of a collection, or ensemble, of particles rather than just one particle. The particles in the ensemble will not all have the same velocity and therefore kinetic energy. However, we should be able to compute an average kinetic energy representative of the ensemble by adding up the kinetic energy of all the particles and dividing by the number of particles in the collection. We will designate this average energy as $<w>$.

As we shall see, when dealing with a collection of particles it is useful to have some measure of the kinetic energy contained in the collection itself, for example the kinetic energy in a certain volume of space at some specific location. We can obtain the total kinetic energy (KE) in each cubic meter (a unit volume) of the collection simply by multiplying the average kinetic energy of the particles by the number of particles in the unit volume n, for example

$$\mathrm{KE} = n<w>.$$

The kinetic energy per unit volume is called the energy density. The units of energy in common use are those for the meter/kilogram/second or MKS system. In this system we call the unit of energy the joule. (One joule is the kinetic energy of a golf ball moving at 6.3 meters or about 19 feet per second.) Therefore, the unit of energy density is joules per cubic meter. For example, the kinetic energy density of the air molecules in your room is about 100,000 joules per cubic meter. Another common unit for energy, particularly in reference to kinetic energy for atomic particles like the electron, is the electron volt. One electron volt is 0.00000000000000000016 joules.

The above equation is useful for describing the kinetic energy in a cubic meter due to the thermal or random motion of the particles contained in that volume. But suppose that the collection of gas also has some non-random motion such that all the particles are moving in the same direction; in other words, flowing somewhere with a velocity V. In this case, we need to add a component to the kinetic energy density that includes the energy resulting from the flow velocity. The mass density ρ, in kilograms per cubic meter, can be expressed as

$$\rho = n\overline{m}$$

where m is the mass of an atom and the bar refers to the *average* mass of the collection of particles. From the general expression for kinetic energy we can see that the flow contribution to kinetic energy density is

$$KE_{flow} = \tfrac{1}{2}\rho V^2.$$

Lets move on to some other types of energy. Force fields such as gravity, electric and magnetic fields also have energy. The energy in a field exists because work has to be done to create the field. Work as defined in physics, the force exerted over a certain distance, is itself a form of energy. For example, if you lift a rock off the ground, you have done work equal to the force required to lift the rock multiplied by the distance it was lifted. In the process the rock gains energy, 'potential' energy. Furthermore, the potential energy the rock gains is exactly equal to the work you did to lift it. The energy the rock has at its new height is called potential energy because the rock has the potential to relinquish that energy by doing new work itself or, if released, to fall toward the Earth and gain—you guessed it—kinetic energy. By the law of conservation of energy the kinetic energy it will gain as it falls the distance it was lifted is exactly equal to the work done and the potential energy gained.

In the example in the previous paragraph, the force field is the Earth's gravitational field. In the case of electric and magnetic fields, charged particles must be moved to create the fields. We can easily create an electric field by separating two opposite charges. Suppose we could reach in and pluck the electron from a neutral hydrogen atom and carry it some distance away. This would leave us with a positive proton and a negative electron separated by some distance. The region between and around the two charges is suddenly filled with an electric field, E. The field looks something like this:

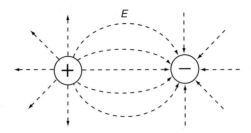

This electric field is a mathematical construction that represents the electric force trying to pull the two charges back together. The field does not exist until we separate the two charges. We had to do work to separate the proton and electron so there must be energy in the electric field that represents this work. The total energy in the electric field equals the work done dragging the electron kicking and screaming away from the proton. Furthermore, if we release the electron it will fly back to the proton gaining kinetic energy equal to the energy in the field and the electric field will disappear. As can be seen by the variation of the density of electric field lines, the strength of the field varies. However, at any point in space, the electric energy in a unit volume of space EE is given by the expression

$$EE = \frac{\varepsilon_0 E^2}{2}$$

where E represents the intensity of the electric field, and ε_0 is a constant that represents the electric conditions in space and makes the units correct to form an equation.

In a similar vein, a moving charge constitutes a current and this current creates a magnetic field surrounding it. This field has the shape of concentric rings around the current. The magnetic field, B, looks like this:

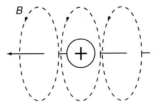

Because work is involved in its creation, the magnetic field also represents a source of energy. At any point in space the magnetic energy ME in a unit volume of space is given by the expression

$$ME = \frac{B^2}{2\mu_0}$$

Like ε_0 in the previous equation, μ_0 is a constant that represents the magnetic conditions in space and that gives the equation the units of joules per cubic meter.

We now have three kinds of energy density; kinetic energy density, electric energy density, and magnetic energy density. If we wanted the total energy in, say, a cubic meter, and if there were particles in motion, and electric and magnetic fields within the cubic meter, we would need to add all three of these. There might be other forms of energy that would need to be included as well.

2 TEMPERATURE AND PRESSURE AND THEIR RELATIONSHIP TO ENERGY

One of the important parameters used to describe a gas (or plasma) is temperature. Except at absolute zero, all atoms (or ions) are in continual random motion. This 'thermal' motion provides the atoms with kinetic energy. In fact, the average kinetic energy $<w>$ of the atoms, molecules, ions or electrons determines the temperature T according to the expression

$$T = \frac{2}{3} \frac{<w>}{k}$$

where k is Boltzmann's constant, a constant that makes the proportionality into an equation with the appropriate units.

It is the continuous thermal motion that makes the atoms (or ions) of a gas (or plasma) collide repeatedly with the walls of a container. This continuous banging creates a force on the walls. The force on a unit area, such as a square meter, is called the pressure. The higher the temperature, the greater the thermal motion, the more the atoms collide with the walls and the higher pressure. Also, the more atoms to do the banging, the greater the pressure. We can conclude that pressure P must be proportional to temperature and also the number of atoms per unit volume that are doing the banging. This can be expressed as an equation:

$$P = nkT.$$

Notice something remarkable. If we substitute our earlier expression for T in the above equation we see that

$$P = \frac{2}{3} n <w>.$$

In other words, pressure and kinetic energy density are the same, apart from a small constant $2/3$,

$$P = \frac{2}{3} KE.$$

3 FORCE AND NEWTON'S SECOND LAW OF MOTION

Alongside energy in importance to physical processes is force. Sir Isaac Newton was the first scientist to quantify the concept of force. He defined force as that entity that can change the motional state of a body whether it is at rest or in motion. That change is called acceleration a. Acceleration is simply the rate of change of velocity or $\Delta V/\Delta t$. He also recognized mass m as the property of a body that tries to resist the acceleration of the body.

Newton's second law of motion is the quantitative embodiment of the meaning of force. The second law states that the sum of the forces acting on a body determines its rate of change of velocity (acceleration) and that that acceleration is inversely proportional to the mass of the body. Unfortunately, the more familiar form of Newton's second law has the mass of the body written on the same side of the equation as the acceleration a:

$$\Sigma F = ma.$$

Since, as indicated earlier, velocity is a vector, acceleration must also be a vector and force F must therefore be a vector as well. It is understood then that the summation of all the forces imposing themselves on an object, indicated here by the Σ, must be achieved by vector addition. It is also important to realize that acceleration must be taken in its most general sense as a change in the *direction* as well as magnitude of the velocity.

Finally, we need to realize that the concept of the 'object' specified by Newton's second law can be generalized to something other than a solid object. For example, a small collection of particles or a parcel of gas or plasma could be considered an object, provided the applied forces act uniformly on all of the particles in the collection. In this case, the mass of the object becomes the sum of all the particles in the collection. In the next section we shall have occasion to talk about the acceleration of a unit volume of gas and our concept of mass density ρ will again be useful.

4 FORMING THE SOLAR WIND

We are now in a position to discuss more precisely the creation of the solar wind. The solar corona cools and becomes less dense as we move away from the Sun. Looking again at our small box of gas seen earlier in Figure 1.2, and sketched below for convenience, we realize that the edge of the box closest to the Sun is bombarded by coronal gas that is hotter and denser than the opposite side. In other words, the pressure on the sunward side is higher than on the side facing away from the Sun.

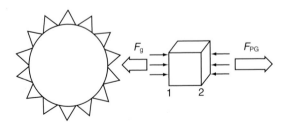

The difference between these two pressures provides a pressure gradient force F_{PG} that pushes the parcel of gas outward, away from the Sun;

$$F_{PG} = \frac{P_2 - P_1}{x_2 - x_1} = \frac{\Delta P}{\Delta x}.$$

The subscripts 1 and 2 refer to the sunward side and opposite side of the box of gas, and $x_2 - x_1$ is the width of the box. The Δs are just a shorthand notation that represents the small difference obtained by the subtraction of the Ps and xs.

An additional force is also acting on the box of gas, the gravitational force of the Sun F_g. The pressure gradient force is the stronger of the two forces and so our parcel of gas will accelerate outward.

Newton's famous second law of motion tells us what acceleration we can expect. This law says the (vector) sum of the forces must equal the mass of gas in the box ρ times its acceleration a.

For our case this becomes

$$\frac{\Delta P}{\Delta x} - F_g = \rho a$$

where the acceleration a tells us how fast the velocity of the box is changing with time, $\Delta V / \Delta t$. This equation is a simplified version of the so-called momentum equation of hydrodynamics and it can be used to describe the acceleration of our little box of coronal gas away from the Sun. Eugene Parker first investigated this equation in 1958. He combined it with an equation that conserves mass and found that a reasonable choice of variables for the coronal temperature and pressure would lead to a supersonic solar wind several solar radii distant from the Sun and that the solar wind would continue to accelerate and achieve a velocity at the orbit of Earth in good agreement with the value later measured by spacecraft.

We can understand the acceleration of the solar wind from an energy perspective by realizing that the acceleration process is a mechanism that converts thermal kinetic energy in the solar corona to flow kinetic energy. Deep in the corona the energy in a cubic meter of coronal plasma is largely thermal kinetic energy. Near the orbit of Earth it is mostly flow kinetic energy.

There is one slight problem, however. Lest we become too smug, nature has a surprise for us. If the foregoing analysis were all there is to it, we would expect the solar wind to cool in exact proportion as it gains flow energy and as it becomes more diluted by expansion. In fact, this is not the case. The solar wind does cool but not as fast as it should as it moves out. Apparently there is another form of energy adding heat to the solar wind as it flows outward. We are not quite sure what this other source of energy is, but there are several types of energy that might do the trick. These include heat carried from the Sun by electrons and special types of waves that travel in plasmas, called MHD waves. A discussion of these energy sources would carry us too far from the main theme of our story.

Things are complicated from yet another quarter. The Sun also has a magnetic field that extends into the corona. Also, our little box of gas is not made up of neutral atoms but rather ions and electrons. This is a plasma and plasmas are usually good conductors of electricity. This means that the coronal magnetic field cannot easily slip through and remain behind as the solar wind plasma departs from the Sun on its outward journey.

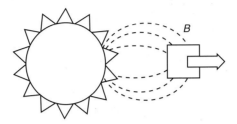

Who will win out? The box of plasma wants to leave because of the pressure gradient from the hot coronal gases but the magnetic field wants to hold it back because the field lines are rooted in the Sun. Here is where the energy density concept comes in handy. Recall the equivalence of energy density and pressure, and the fact that pressure is just force per unit area. Using this, to see who wins, we simply compute the magnetic energy density and the thermal kinetic energy density. At the hottest point in the corona, the thermal kinetic-energy density turns out to be about twice as great as the magnetic energy density so the magnetic field is dragged helplessly away from the Sun.

Even though the magnetic field effect on the motion of our box of gas is small, we should correct our momentum equation to include the magnetic field. The magnetic field enters in two ways.

First, since the magnetic field represents energy, and we have seen that energy density translates into pressure, we need to add the magnetic pressure to the gas pressure in the pressure gradient force term in the momentum equation. The total pressure is now

$$P_{\text{total}} = P_{\text{gas}} + P_{\text{magnetic}} = nkT + \frac{B^2}{2\mu_0}.$$

Secondly, there is a real magnetic force holding back the box of plasma. This force needs to be added to the two forces we already know about. As the plasma tries to move outward, the magnetic field generates currents within the plasma just as a magnet in an electric generator generates currents in its moving wires. As we shall see in the next section, a current flowing in a magnetic field experiences a force. This force is known as the $J \times B$ force. It is proportional to the density of the current J and the magnetic field intensity, and it is always at right angles to both. We will simply call it the magnetic force F_{m}. Our corrected momentum equation now becomes

$$\frac{\Delta P_{\text{total}}}{\Delta x} - F_{\text{m}} - F_{\text{g}} = \rho a.$$

Because it includes the effects of the magnetic field, this is a simplified form of the momentum equation of *magnetohydrodynamics*, or MHD for short. It is a very useful equation, particularly when combined with a similar equation for energy and one for the conservation of mass. This equation is not written in the most general form. It would normally be written in the language of calculus, and also we have assumed that the density of the gas or plasma remains unchanged, which of course is not correct for very long in an expanding plasma like the solar wind.

This is also the fundamental equation for determining convection in the magnetosphere, which you will recall, like convection in our open convertible, is driven by pressure gradient forces. Great mathematical models of the magnetosphere are built around the solution of this equation: the so-called MHD models. These models offer the best hope for precise simulation and forecasting of storms in space.

5 MAGNETIC FIELDS, MAGNETIC FORCE AND MAGNETIC MERGING

We have seen how a charge in motion generates a magnetic field around it and how this field has the general shape of concentric circles, and we have hinted that a magnetic field exerts a force on a current. We have also seen that the magnetic field represents a form of energy with a certain energy density. Since the motion of charges creates a magnetic field, it makes sense that if a magnetic field disappears we might see the missing magnetic energy appear in the form of accelerated charged particles.

In space there are no metal wires to confine moving charged particles to a linear current, but is helpful to use wires to visualize some important effects. Imagine that you are holding two long parallel wires side by side. Both carry the same amount of current flowing in the same direction. If the current is strong enough and the wires close enough together you will actually feel a force trying to pull the wires together, just as if you were holding two magnets—which, in fact you are. If you now try to pull the wires farther apart you will have to do work against the magnetic force trying to pull the wires together. This work goes into potential energy as a magnetic field builds up between the two wires. Sounds a lot like separating our positive and negative charges earlier. The similarity does not stop. Suppose you release the two wires. They will fly back together again gaining kinetic energy in the process. The magnetic field built up in the space between the wires disappears and its energy reappears as kinetic energy as the wires slam together. You have just experienced magnetic field annihilation or 'magnetic merging' as it is more commonly called.

Magnetic merging in the context of space weather does not involve wires, of course, but the motion of charged particles controlled by forces such as we discussed in the last section produces large-scale currents in space. Currents deep within the Earth and in the magnetosphere shape the magnetic field in the magnetosphere and at the day-side magnetopause. Currents within the Sun and the solar wind shape the magnetic field carried toward the Earth by the solar wind.

Specifically, in a magnetic cloud responsible for a magnetospheric storm a combination of plasma currents flowing perpendicular to the ecliptic plane and currents moving along the magnetic field result in a helical-shaped field with strong north–south components.

The currents carried by the solar wind rushing toward the currents of the Earth's magnetosphere can result in the annihilation of magnetic energy at the magnetopause. This magnetic energy must reappear as kinetic energy of accelerated ions and electrons that make their way into the magnetosphere through various paths and ultimately contribute to the severity of the magnetospheric storm in the variety of ways we have witnessed.

6 SINGLE PARTICLE MOTION

Most of our discussion up to now has treated the particles in space as belonging to the collection of particles we call the plasma. This approach works well as long as the individual particle energies are low and can be represented by the temperature of the plasma. However, some processes selectively accelerate individual particles. As their energy grows, these particles begin to interact independently with the electric and magnetic fields. It is then necessary to use different physical techniques to analyze their behavior. We need to look at the behavior of individual charged particles in magnetic and electric fields. Newton's second law will again help us, but the object experiencing the forces will now be single charged particles rather than a collection of particles.

The effect of an electric field on a charged particle, like a proton or electron, is easy to visualize because the force always acts along the field. For a positively charged particle like a proton, the force is in the same direction as the field, and for a negative particle, like an electron, it is in the opposite direction. Furthermore, the electric force is just the product of the electric field strength E and the particle charge q:

$$F_{\text{electric}} = qE.$$

The force on a charged particle in a magnetic field is a bit more complicated. The magnetic force only acts on a moving charge, and furthermore it always acts at right angles to the direction of motion. As can be seen in the three-dimensional figure below, this means that the path of electrons or protons in a uniform magnetic field is a circle in the plane perpendicular to the magnetic field. Protons move clockwise and electrons counterclockwise in an upward-pointing field.

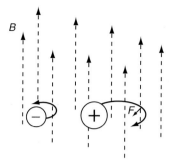

If the electron or proton also has some velocity in the direction of the magnetic field, then the particle motion will be a helix whose axis lies along the field, as shown below.

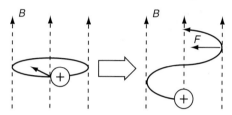

Next, imagine what will happen if the magnetic field is no longer uniform but instead converges. The magnetic force still remains perpendicular to the magnetic field. This produces a component of the force away from the direction of convergence with the result that the helix is forced to flatten out. At some point the particle's spiral will actually reverse direction. The particle is said to have 'mirrored' or 'bounced'.

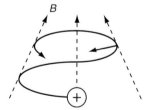

In the Earth's magnetic field the field lines have a natural convergence as they curve toward the Earth in each hemisphere. This provides a trap for charged particles as they mirror and bounce back and forth between the converging magnetic field in each hemisphere. We now see how the Earth's magnetic field provides a magnetic 'bottle' that holds the higher energy Van Allen radiation belts and lower energy particles that form the ring current. Because there is very little atmosphere at the high altitude where these particles bounce repeatedly from hemisphere to hemisphere, there is nothing to slow them and they can remain for very long periods of time.

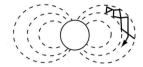

What we have discussed to this point is only a *qualitative* description of the motion of charged particles in electric or magnetic fields. At some point in a complete description of the magnetosphere it becomes necessary to know the path of a particle more precisely. We can rely on Newton's second law, with the correct force inserted, to help us once again. Using the methods of calculus, also devised by the ingenious Newton, it is possible to compute the exact position of a particle at each instant of time. Unfortunately, this is a tedious and time consuming calculation, even for supercomputers. The problem is made more complicated because the number of particles is astronomical. We can compute sample trajectories for a few test particles to get a general idea of what might be going on, but to make real progress with the big picture we have to find a more expedient approach.

7 THE GUIDING CENTER APPROXIMATION

Often in science great progress can be made by using approximations. A good example is the 'guiding center' approximation for the motion of charged particles under the influence of magnetic and electric fields. We have seen that a charged particle moving perpendicular to a uniform magnetic field will move in a circle. The center of that circle can be thought of as an 'approximation' for the position of the particle so long as we are only concerned with what happens over a time that is long compared with the period of one orbit—sort of a time-average position. For most situations this time-average is good enough, so the position of the center of the circle is a good enough approximation of the position of the particle. It is called the guiding center.

Suppose that the magnetic field is not uniform but changes in some direction, or that there is also an electric field present. These additional conditions will cause the particle's circular orbit to have a larger radius of curvature at some point so that the particle will execute quasi-circular or cycloidal motion. It will drift slowly in one direction. We can still define a guiding center but that guiding center will now move in the direction of the particle drift. The guiding center has a velocity and that velocity is a good approximation of the time-average behavior of the particle.

If we had to compute the exact position of the particle to determine the motion of the guiding center there would not be much advantage to defining and using the guiding center approximation. Fortunately, that is not necessary. It is possible to derive a set of equations that give us just the motion of the guiding center for three important kinds of situations: the case where the magnetic field changes magnitude in a direction perpendicular to the direction of the field; the case where there is some curvature to the magnetic field; and finally, the case where there is an electric field present (shown below).

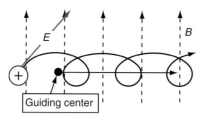

Guiding center

The guiding center equations depend on certain properties of the charged particle, its orbit and the magnetic and electric field. They can be used to build a model to trace the motion of electrons and ions in the magnetic and electric fields of the magnetosphere and are the basis of a class of magnetospheric models. These models complement MHD models because they give the approximate motion of individual particles. The Magnetospheric Specification Model is one model that uses the guiding center approximation.

8 ADIABATIC INVARIANTS

There is another set of approximations that help us determine the behavior of particles in the magnetosphere. Certain properties of particle orbits don't change very much if changes in the magnetic field are small over time scales related to the orbit and if energy from outside is not added to the particles. These properties are called adiabatic invariants; adiabatic because input energy is not considered important, and invariant because the properties are approximately constant during the motion.

There are three important adiabatic invariants in magnetospheric physics. The first of these confirms our earlier claim that particles spiraling about the Earth's magnetic field lines will bounce and be reflected back up as they move downward toward the stronger magnetic field near the Earth. In the language of adiabatic invariants, this mirroring behavior occurs because the 'magnetic moment' of the particle must remain nearly constant as the charged particle spirals downward in a tighter spiral. For this to happen the pitch of the helix must increase until the helix is flat and then the pitch must reverse—the particle reverses direction along the field. This adiabatic invariant is very useful because it dictates the strength of the magnetic field at which the particles of a given pitch will mirror.

The second and third adiabatic invariants are more difficult to explain in simple terms but they provide information on the way in which particles bounce from hemisphere to hemisphere and drift around the Earth.

Adiabatic invariants have played a very large role in our current understanding of storms in space. They make it possible to provide a good description of the Van Allen radiation belts without understanding the origin of the particles themselves.

9 SUMMARY

Space is gradually yielding its weather secrets. Our greatest tools in this effort are powerful computer–mathematical models that simulate the space environment with continually increasing accuracy along with the continued exploration of space by increasingly sophisticated spacecraft. Together these two powerful technologies are our path to preventing the range of problems associated with storms in space.

FURTHER READING

Further discussion of the physics of storms can be found in the following books:

Kallenrode, May-Britt (1998) *Space Physics: An Introduction to Plasmas and Particles in the Heliosphere and Magnetosphere,* Springer Berlin Heidelberg.

Parks, George K. (1991) *Physics of Space Plasmas—An Introduction,* Addison Wesley, Redwood City, CA.

Suess, Steven T. and Bruce T. Tsurutani (eds) (1998) *From the Sun: Auroras, Magnetic Storms, Solar Flares, Cosmic Rays.* American Geophysical Union, Washington, DC.

Glossary of Space Weather Terms and Acronyms

ACE Acronym for the Advanced Composition Explorer spacecraft. ACE is stationed at the L1 point (see below) and provides important data on the solar wind before it hits the magnetosphere.

AE A magnetic index that measures the total magnetic distortion of the Earth's magnetic field as a result of electric currents flowing in the ionosphere associated with the aurora.

AL A magnetic index that measures the negative magnetic distortion of the Earth's magnetic field as a result of electric currents flowing in the ionosphere associated with the aurora.

Artificial neural network An artificial neural network is the computer equivalent of the biological neural network in the human brain. Just as for biological networks, artificial neural networks can be trained to 'learn' relationships and even forecast events.

Bow shock Because the solar wind speed exceeds the magnetohydrodynamic wave speed, the solar wind is supersonic as it passes the Earth. This gives rise to a bow shock wave at the nose of the magnetosphere. The bow shock slows the solar wind and converts the high energy of the solar wind flow into heat.

COMSAT A contraction for communication satellite. Also see geostationary below.

Corona (of the Sun) The outer layer of the Sun's atmosphere where the temperature rises to several million degrees. The corona is visible only during an eclipse and is the region of the solar atmosphere that gives rise to the solar wind.

Coronal mass ejection or CME An outburst of coronal gas that becomes a cloud of dense plasma and enhanced magnetic field. CMEs are an important cause of magnetospheric storms.

Dipole (magnetic) The shape of an undistorted magnetic field produced by a loop of current. The Earth's magnetic field would be a dipole if it were not for the solar wind.

Dst An acronym for 'Disturbance storm', an index computed from ground magnetometer data that indicates the intensity of a magnetic storm. Dst is largely a measure of the ring current.

Electron An elementary atomic particle that orbits the atomic nucleus and is important in determining the chemistry of the elements. In space, electrons can exist independently as free particles.

Electron volt A unit of energy popular with physicists who work with atomic particles.

EVA Extra-vehicular activity: that portion of an astronaut's activity which takes place outside the spacecraft.

Flare A solar flare is a region on the Sun characterized by an intense release of energy resulting in the emission of radiation and often energetic charged particles. Flares may last a short time or they may last for hours. They can be seen in the H alpha wavelength and by the X-rays they emit. The energy from flares is believed to come from the annihilation of magnetic fields.

Flux Flux may be used to describe the intensity of a flow of particles: the number of particles incident in a specified area in a given time, for example the number of electrons crossing a square meter of space in one second.

Geostationary (orbit) The name given to the circular orbit at an altitude of about 36,000 kilometers near the equatorial plane of the Earth. The orbit speed of a satellite in this orbit matches the rotation of the Earth so the satellite appears to remain suspended above one location on the Earth. An antenna on the Earth can remain pointed at the satellite without being moved as the Earth rotates. This orbit is particularly useful for communication, weather, and surveillance satellites.

GOES An acronym for a series of NOAA satellites: the Geostationary Orbit Environmental Satellites. These satellites provide weather and space environmental data.

GPS Global Positioning System. A fleet of satellites which broadcast radio signals that allow precise determination of a user's location (and altitude) on the Earth.

H alpha The designation for an emission line of neutral hydrogen at 656.3 nanometers used for observing active regions on the Sun.

IMF An acronym for the interplanetary magnetic field (see below).

Interplanetary magnetic field or IMF The magnetic field of the Sun is drawn out into interplanetary space by the motion of the solar wind. It is referred to as the interplanetary magnetic field or IMF.

Ion When an atom loses one or more of its electrons, it becomes an ion. The ion is then positively charged to some multiple of the charge on an electron. Atoms can also gain electrons in space to become negative ions but this is not an important effect for most magnetospheric processes.

Ionization The process of creating an ion by removal of an electron.

Ionosphere The upper region of the Earth's atmosphere is exposed to ionizing electromagnetic radiation from the Sun and energetic particles precipitating downward from the magnetosphere. These waves and particles have sufficient energy to knock electrons from the neutral atoms of the atmosphere and create a layer of electrons and ions at high altitudes called the ionosphere. The charged particles in the ionosphere can carry electric currents (ionospheric currents) that can produce important magnetic fields felt on Earth. These currents are an important aspect of a magnetospheric storm. Also, the number of electrons in the ionosphere controls the propagation of radio waves in some important frequency ranges.

Kiloelectron volts (keV) A kiloelectron volt is equal to 1,000 electron volts (see above).

Kp Kp is the name for an index of magnetospheric storm activity. It is computed from data from magnetic observatories around the world. It is a semi-logarithmic scale from 0 to 9 somewhat like the well-known Richter scale for earthquakes.

L1 L1 is a point of equilibrium in the Sun–Earth gravitational system that allows a spacecraft to remain suspended in a stable orbit on a line between the Earth and Sun. It is $\frac{1}{100}$ of the distance to the Sun from the Earth.

Limb The edge of the Sun as seen from the side. Coronal mass ejections (see above) and prominences can be seen best as they emerge from the limb and can be seen in profile.

Magnetic cloud A magnetic cloud is a region where the quiet solar wind is disturbed by the outflow of enhanced plasma and magnetic field from an eruption on the Sun (often a CME). Magnetic clouds carry a particular twisted configuration of magnetic field that can interact strongly with the magnetosphere and produce large geomagnetic storms. Magnetic clouds can create shock fronts as they push through the slower moving solar wind.

Magnetohydrodynamics Magnetohydrodynamics is the branch of physics that treats a plasma with an embedded magnetic field as a continuous fluid.

Magnetometer A magnetometer is an instrument for measuring magnetic fields.

Magnetopause The outer surface of the magnetosphere is called the magnetopause. Many important processes that couple the energy in the solar wind to the magnetosphere take place at the magnetopause.

Magnetosphere The magnetosphere is the region surrounding the Earth where the motion of charged particles is controlled mainly by the magnetic field of the Earth. The flow of the solar wind past the Earth influences the behavior of magnetospheric plasma and causes great variations in the motion and quantity of the particles within the magnetosphere. Large enhancements in the intensity and motion of these particles can alter the configuration of the magnetosphere giving rise to magnetospheric (aka. geomagnetic) storms. The magnetosphere extends outward from the Earth about 60,000 km toward the Sun and has a long tail that extends many times that distance in the direction away from the Sun. A bow shock wave forms in the supersonic solar wind just upstream of the nose of the magnetosphere.

Magnetospheric Specification Model (MSM) The MSM is a mathematical computer model that can simulate the intensity of magnetospheric storms by computing the fluxes of electrons and ions that will be felt by satellites. It uses data obtained in real-time from spacecraft located upstream of the Earth and also data from ground-based magnetometers. The MSM is a first-generation model for the prediction of storms in space. It is analogous to computer models used to forecast the intensity of storms on Earth. The MSM is used by the National Oceanic and Atmospheric Administration (NOAA) and the US Air Force.

Magnetospheric storm A magnetospheric storm is a disturbance within the Earth's magnetosphere associated with changes in magnetic fields, enhanced fluxes of plasma, increased intensity of the Van Allen radiation belts (see below), and increased auroral activity.

Magnetotail The magnetotail is the long tail of the magnetosphere that extends down(solar)wind like the trailing end of a windsock. The magnetotail is the region where some of the energy derived from the interaction between the solar wind and the magnetosphere may be temporarily stored before it is released by the earthward motion of plasma in a magnetic storm or substorm.

MHD MHD is the acronym for magnetohydrodynamics.

Model In our context a model is a computer program that simulates a real physical situation by mathematical calculations. A model has data inputs that set the conditions for the model. The model output is the desired simulation which provides the calculated values at many desired locations in space.

Neural network Same as artificial neural network (see above).

Neutron A neutral elementary particle that is a component of the nucleus of all atoms except hydrogen. The neutron is unstable outside the atom. It quickly undergoes radioactive decay to form a proton, an electron, and an anti-neutrino.

NOAA The National Oceanic and Atmospheric Administration.

Photosphere The photosphere is the layer of the atmosphere of the Sun that is the yellow disk visible to the human eye. (Never look directly at the Sun.) The photosphere lies below the corona and has a temperature of 5,800 kelvin.

Plasma Plasma, as used in physics, refers to a gas that is ionized and may possess an embedded magnetic field.

Plasmasheet A layer of dense plasma that lies at the middle of the magnetotail where the magnetic field is weak. The plasmasheet is the region from which protons and electrons enter the magnetosphere on the night side during a magnetospheric storm or substorm.

Proton A positively charged elementary particle that is a component of the nucleus of atoms but which can exist as a free particle outside the nucleus.

Riometer An instrument for measuring the number of electrons in the ionosphere.

SOHO SOHO is an acronym for Solar and Heliospheric Observatory. It is a spacecraft parked at the L1 point where it can continuously image the Sun and measure X-rays and particles from the Sun.

Solar energetic particles or SEPs Solar energetic particles are particles accelerated by the outward motion of CMEs or in solar flares. They are mainly protons and electrons. Other ions are present in lower numbers. The importance of SEPs lies in their high energy, which allows them to penetrate surfaces and do damage through the ionization of materials or even human flesh and organs.

Substorm A substorm is a period of increased auroral zone activity lasting about an hour. It is associated with the impulsive injection of fresh magnetospheric particles from the magnetotail.

Van Allen radiation belts The Van Allen belts are two zones of intense protons and electrons trapped within the magnetosphere. The inner zone lies close to the Earth and the outer zone lies beyond it and extends nearly to the magnetopause.

Vector A vector is any quantity that possesses both magnitude and direction. Wind velocity is an example of a vector.

References

Allen, J.H. and D.C. Wilkinson (1993) Solar-Terrestrial Actvity Affecting Systems in Space and on Earth, in J. Hurska, M.A. Shea, D.F. Smart, and G. Heckmann (eds), *Solar-Terrestrial Predicitions – IV*, Vol. 1, NOAA, Boulder, CO

Blais, G. and P. Metsa (1993) Operating the Hydro-Quebec Grid Under Storm Conditions Since the Storm of 13 March, 1989, in J. Hurska, M.A. Shea, D.F. Smart, and G. Heckmann (eds), *Solar-Terrestrial Predicitions – IV*, Vol. 1, NOAA, Boulder, CO

Committee on Solar and Space Physics, Space Studies Board, National Academy of Science (2000) *Radiation and the International Space Station*, National Academy Press, 2000

Kappenman, J.G. (1993) Geomagnetic Disturbances and Power System Effects, in J. Hurska, M.A. Shea, D.F. Smart, and G. Heckmann (eds), *Solar-Terrestrial Predictions – IV*, Vol. 1, NOAA, Boulder, CO

Kleusberg, A. (1993) The Global Positioning System and Ionospheric Conditions, in J. Hurska, M.A. Shea, D.F. Smart, and G. Heckmann (eds), *Solar-Terrestrial Predicitions – IV*, Vol. 1, NOAA, Boulder, CO

Kunstadter, C.T.W. (1999) Were We Crying 'Wolf', in Insurance Implications of the Leonids and Other Space Phenomena, *Space Weather Week*, NOAA, SEC, Boulder, CO, May 13, 1999

Lanzerotti, L.J. (1983) Geomagnetic Induction Effects in Ground-based Systems, *Space Science Reviews*, Vol. 34

Lett, J.T., W. Atwell, and M.J. Golightly (1990) Radiation Hazards to Humans in Deep Space: A Summary with Special Reference to Large Solar Particle Events, in R.J. Thompson, D.G. Cole, P.J. Wilkinson, M.A. Shea, D. Smart, and G. Heckman (eds), *Solar-Terrestrial Predictions: Proceedings of a Workshop at Leura, Australia, October 16–20, 1989*, NOAA, Boulder, CO

Maynard, N. (1995) Space Weather Prediction, *Reviews of Geophysics*, July 1995

Pratt, R.L. (1993) Geomagnetic Effects on Manufacturing Processes, in J. Hurska, M.A. Shea, D.F. Smart, and G. Heckmann (eds), *Solar-Terrestrial Predicitions – IV*, Vol. 1, NOAA, Boulder, CO

Acknowledgements

I am indebted to Alex Dessler for suggesting this book, for his continuing encouragement throughout its preparation, and for reading and making valuable suggestions on an early draft. I am also grateful to George Siscoe for reading the manuscript and offering suggestions and corrections, and for writing the foreword.

Mary Armeniades, a dear friend and fellow author, read a draft and provided useful comments from the standpoint of a non-scientist. I am particularly indebted to Joe Allen who was extremely generous with his time in agreeing to his interview and in taking the time to proof my transcription.

My colleagues at Rice University, faculty, staff, and students too numerous to name, through their collegiality and dedication to space research, have provided an ideal environment for the development of this book. Moreover, the space physics group at Rice University deserves credit for shaping some of the knowledge of the magnetosphere represented in this book. My interest in space was first sparked by the enthusiasm and genius of my early mentor at the University of Iowa, James A. Van Allen.

The staff at Cambridge University Press, particularly Jacqueline Garget, Mairi Sutherland, and Dr. Matt Lloyd have been very professional and a real pleasure to work with.

Finally, my wonderful wife has exhibited substantial patience with my long hours and has provided inspiration, support and advice when it was most needed.

Grateful acknowledgement is made to the following for permission to reprint figures or use previously published material:

Dan Urbanski of the Silver Image Studio for the cover photo;
Yohsuke Kamide for the use of his well-known drawing of the magnetosphere;
Lou Frank and John Sigwarth of the University of Iowa for the use of images from the POLAR spacecraft VIS imager;
The High Altitude Observatory and The National Center for Atmospheric Research for the use of their solar image slide collection;
The SOHO EIT and LASCO science teams for images of the Sun;
Greg Byrne and NASA space shuttle astronauts for the picture of the aurora taken from the Space Shuttle;
NASA and the NOAA centers: The National Geophysical Data Center and The Space Environment Center for several figures;

Index